From Daladala to DART: Transforming Dar es Salaam's Public Transport

Sanjay

LIST OF CONTENTS

CHAPTER ONE

1.1 Background

1.1.1 URBAN TRANSPORT IN DAR ES SALAAM

Dar es Salaam is one of the most rapidly growing cities in Africa (SUMATRA, 2011). Before the first phase of the Bus Rapid Transit (BRT), specifically, the Dar es Salaam Rapid Transit (DART), city commuters commuted in all parts of the city by privately owned buses commonly known as "Daladala" in Kiswahili (Figure 1). There were about 5,200 mid-sized buses with the capacity of 40 passengers (DART, 2014). *Daladalas* were estimated to be providing three million trips on a given working day. However, these bus services were not properly co-ordinated, neither were they based on standardised facilities. As a result, the Daladala system was characterised by inefficiency, inconvenience, unreliability, and discomfort. Therefore, this trend necessitated the establishment of an organised dependable bus transport system to serve city residents. In order to get rid of the snags inherent in the urban transport and to improve living conditions of city residents, the government embarked on the improvement of the urban transport by providing a reliable and efficient public transportation system. BRT was chosen as a mode as the system is based on buses (Figure 2) which require lower capital investment than the rail; in addition, it is flexible (Moylan, Clonts, Wang, O'Connor, & Lounsbury, 2013). The system's operations have been successful in other parts of the world, particularly in the cities such as Lagos, Sao Paulo, Johannesburg, Curitiba, Bogota, Beijing, Los Angeles, and Taipei (Ugo, 2014). In fact, the successful and functional implementation of the system can serve as a quality mass transit (Adebambo & Adebayo, 2009) and meet travel needs of city commuters at affordable rates (UITP 2010).

1.1.2 DART IMPLEMENTATION

In the efforts of meeting its urban transport demand, the city of Dar es Salaam launched the first phase of DART in May 2016 (Citiscope, 2017). This is part of the planned 130km system aimed at catering for 90 per cent of the city commuting population expected to have grown to five million people by 2015 (Figure 3). At the time of this study, DART had been in operation for 24 months. Its first phase (Figure 4) started by covering 20.9km of trunk lanes, 57.9 km of feeder routes, 27 stations

1

(23 with overtaking lanes, and 4 without overtaking lanes), five terminals and seven feeder stations (Figure 5). Initial demand was estimated at about 410,000 passengers per day translating into total annual ridership of 130 million trips (DART, 2014).

Figure 1: *A typical daladala bus. Source: Researcher*

Figure 2: *A typical DART bus. Source: Researcher*

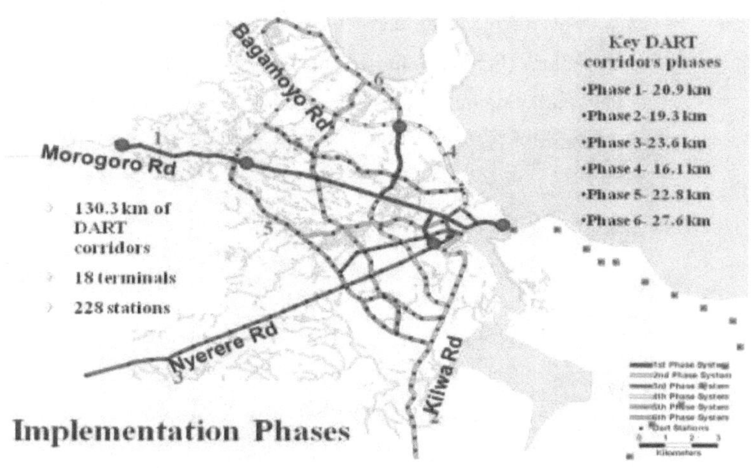

Figure 3: *DART system map with implementation phases: Source DART project information memorandum (2014)*

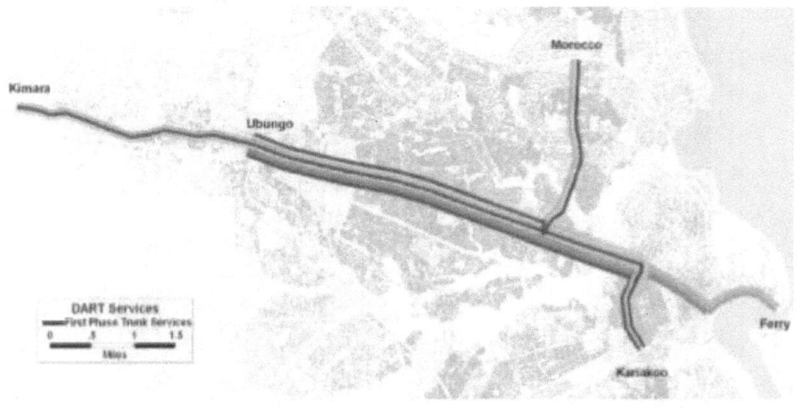

Figure 4: *Phase I Trunk services map: Source DART project information memorandum (2014)*

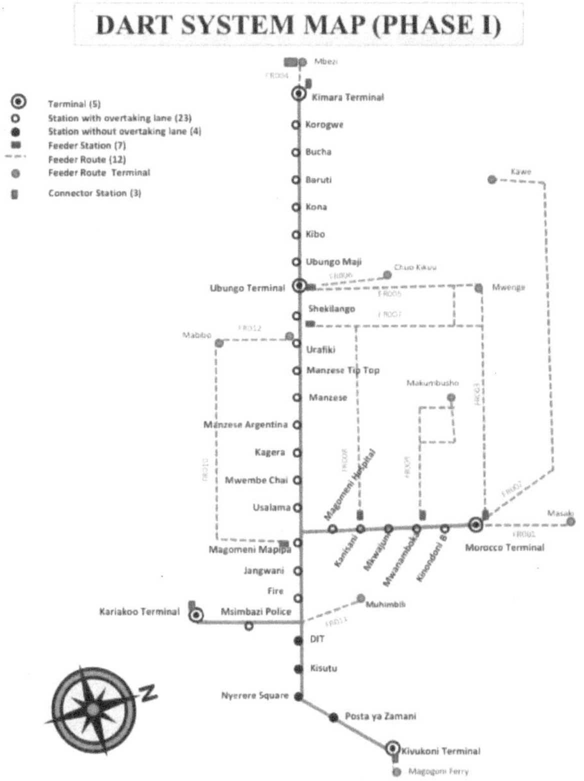

Figure 5: *DART Phase I network: Source DART project information memorandum (2014)*

1.2 Problem statement

The DART project seeks to address the shortcomings of the existing Daladala system and provide a better public transport alternative to Dar es Salaam city residents. Thus, the system needs to address fully problems related to the Daladala system. In this regard, DART should not just be a replacement of the Daladala system, but it should also eliminate—to a large extent—all problems associated with the Daladala system for it to command a wider usage. Wider usage would assist city residents in reducing travel time when moving from one point to another in various parts of the city, in addition to reducing the use of private vehicles in Dar es Salaam (Ugo, 2014). As Ugo (2014) observes, providing quality service and meeting customer satisfaction pose a challenge to any system that intends to provide effective urban transport. It is, therefore, important to assess the customer satisfaction of the existing urban transport system.

Overall, any shortcomings that are manifested in DART's operations in its first phase constitute a problem, which needs addressing to improve the current phase. Such adjustments are certainly going to make future phases be designed such that the existing shortcomings are either reduced or eliminated. As DART is a new system to many Dar es Salaam city residents and especially to users of public transport system, several snags have emerged during its early days. Solutions and rectification of such snags are necessary for the system to attract a wider usage of both captive passengers who are concerned with satisfaction related to reliability, comfort and cleanliness (Deng & Nelson, 2012) and transit choice riders. This would ultimately reduce traffic congestion, as passengers and riders would be more attracted by shorter travel times when using DART than is the case when using private cars (Cain & Flynn, 2013).

This study assessed the DART users' satisfaction. It has identified the existing shortcomings and their causes. Gathered information and the study recommendations would be useful to the relevant government authorities and other stakeholders in making informed decisions about city commuting system generally and in scaling up of the DART operations system in particular.

1.3 Objectives of the study

1.3.1 GENERAL OBJECTIVES

The general objective of the current study was investigating how the current DART services are delivered with a view of improving urban transport in the city of Dar as Salaam.

1.3.2 SPECIFIC OBJECTIVES

1. To find out how customers (commuters) perceive the existing service quality attributes

2. To establish shortfalls based on customers' viewpoints and suggest measures to be taken for service improvement.

3. To establish whether adequate communication approaches were used as incorporation of public opinion increases the chances of project success as the public may understand the project better if involved. (Kumar, Zimmerman & Agarwal 2012).

1.4 Purpose and rationale of the book

1.4.1 PURPOSE OF THE BOOK

The purpose of this book is to fulfil the requirements for the completion of a master's degree at the University of Cape Town. However, the information gathered would be made available to the government of the United Republic of Tanzania, and it may help the government in making necessary adjustments in the subsequent phases of the envisaged expansion of DART network. The current study is also intended to generate insights for the smooth and efficient operation of DART services that would serve the Dar es Salaam community better. The information gathered from passengers using DART has informed this study and would ultimately inform stakeholders involved in future phases of DART with the expectation that they would make sound decisions during both the design and construction stages. The ultimate result would be value for money for the investment made by the government and, hence, generate higher satisfaction on the part of DART users.

1.4.2 RATIONALE OF THE STUDY

The study rationale is centred on the reality that customers' views play a significant role in evaluating the quality of transit services. This is because customers are the real consumers of the services (Nkurunziza, Zuidgeest, Brussel, & Bosch, 2012a). Since public transport service quality plays an important role in shaping daily transport habits of urban residents (Hussein & Hapsari 2014), improving service quality is important and can be achieved better by assessing the existing service quality.

1.5 Research questions

1. What is the users' overall satisfaction level with DART services?

2. What is DART passengers' acceptance level of the system?

3. To what extent is the service affordable?

4. What challenges do BRT users face when using the system on day-to-day basis?

5. Was the level of public participation (the public being informed and giving views or opinions on BRT) while implementing the completed phase of the project adequate?

6. Does BRT infrastructure provide universal access?

7. Are the buses operators use in phase one appropriate?

CHAPTER TWO

2.0 LITERATURE REVIEW

2.1 Introduction

This chapter provides the background information of the study that facilitated the survey instrument formulation. Moreover, it identifies a list of factors that affect passenger satisfaction. The factors were selected and then translated into a practical list that was used in the questionnaire. Furthermore, the chapter identifies and discusses some analytical methods used and not used in the study before delineating the method applied in addressing the study objective.

2.2 Conceptual definitions

This sub-section contains a list of definitions and terminologies used or referred to in various chapters of the book. These terminologies include:

2.2.1 FEEDER STATIONS

These are stations found along the off trunk roads and placed near trunk stations. They facilitate the transfer between feeder routes and trunk stations (Figure 6).

2.2.2 TRUNK STATIONS

These are stations built along the trunk routes. They consist of passenger platforms, and accessed via the pedestrian zebra crossings. The platform height guarantees pedestrian safety (Figure 6).

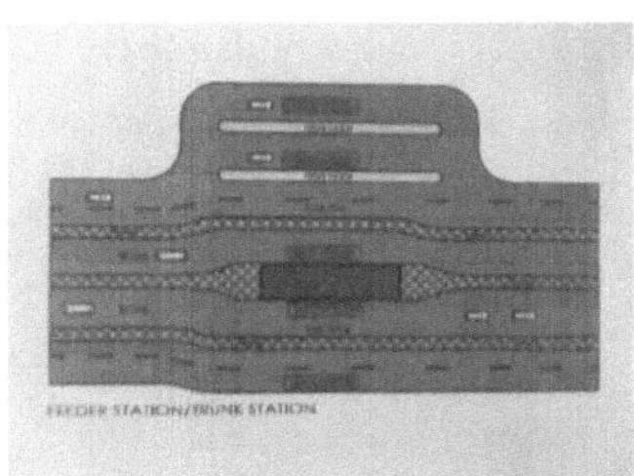

Figure 6: *Trunk/Feeder station layout. Source: Researcher observation*

2.2.3 FEEDER ROUTES

These routes connect feeder stations to different parts of the city (Figure 7).

Figure 7: *Example of a feeder station and feeder route (Shekilango Station). Source: DART project information memorandum (2014)*

2.2.4 TERMINALS

These are large stations located at the start and end of trunk route roads or in between for buses to turn around. In the first phase of DART, end terminals have been located at Kivukoni, Kimara, Morocco, and Kariakoo. Ubungo is the in-between terminal. Terminals facilitate transfer between feeder routes in addition to giving access to other transportation facilities including private vehicles (Figure 8).

Figure 8: *Layout showing a connector station and a terminal (Kivukoni). Source: DART Project information memorandum (2014)*

2.2.5 FEEDER ROUTE TERMINALS

These are terminals at the start of feeder routes.

2.2.6 CONNECTOR STATIONS

These are large feeder stations with multi-terminal connectivity (Figure 8).

2.3 Urban mass transportation experience from other countries

Generally, people living in urban areas make trips from one point to another because of activities they undertake or when on vacation. These activities are related to time and place (Amiegbebhor & Dickson, 2014). Transportation in urban areas is, therefore, necessary to support people's activities. As these activities are related to time and place, the efficiency of activities relies on effective time management and arrivals at destinations within the desired time. City commuters in many parts of the world waste a lot of time moving from one point to another and, thus, making many of their economic activities largely inefficient. In consequence, there has been a loss of income and, ultimately, reduction of the economic performance of city dwellers.

To address this problem, efforts have been made in various cities to improve public transportation. BRT has been one of the modes expected to offer improved transportation to city dwellers in various cities as the system reduces travel time, which influences ridership and reduces usage of private vehicles, and hence reducing fuel consumption and greenhouse gas emissions (Moylan *et al.*, 2013). From 2009 onwards, many cities in Africa such as Johannesburg, Cape Town, Dar es Salaam, Marrakech, and Lagos have been implementing this mode of transportation. These projects are usually executed in phases due to the high capital required to complete all the sectors within a particular city. The modes of funding for these projects vary from one city to another. In Dar es Salaam, the project was financed by the African Development Bank (ADB), World Bank (WB), and the Government of Tanzania (GoT) (World Bank, 2013) whereas in Johannesburg it was financed by the National Public Transport Infrastructure and Systems Grant (Allen, 2013).

2.4 Assessment of public transport quality

Public transport services can be improved through policies and strategies that come out through evaluation of current level of service provision (Morton, Caulfield & Anable, 2016). The assessment of quality of public transport is a common practice in various cities around the world and it aims at bringing about reforms and improves quality.

Potogvkina and Ugay (2013) assessed the quality of service of passengers in Vladivostok, Russia. The researchers used passenger survey as a method of sourcing information.

Olikova (2016) evaluated the quality of public transport in terms of passenger satisfaction. The study was carried out for public transport system in the city of Ostrava, Czech Republic. In this study, information was sourced by asking passengers to fill in questionnaires. A five-point scale was used to determine the level of satisfaction of a given criterion. The evaluation criteria consisted of

- *Domain defining for the utility function.*
- *Graphical representation of values using dot diagram.*
- *The use of least square method to determine regression functions type.*
- *Division of partial criterion utility function's definition domain into nominal values intervals and determination of limit nominal values*

Morton *et. al.* (2016) examined the quality of service indicators used by the Scottish government to assess opinions of passengers towards bus transport. A bespoke quality of service indicator is used by Scottish government to monitor perceived experiences of passengers using local buses. It comprises of multi-item attitudinal scale which allow passengers to express their experience on the opinion statements (scale items) on a five point Likert-scale which ranges from highly disagree to agree. In addition, a single-item scale was used to evaluate the perceived level of satisfaction of public transport.

In reviewing a study by Redman, Friman, Garling and Hartig, (2013), Morton *et. al.* (2016) have indicated the existence of research concerning transport sector quality as briefly outlined in this section and summarized in Table 1.

Morton *et al.* (ibid) have reviewed a work by Fick and Ritchie (1991) in which SERVQUAL was used. The authors have illustrated how the measurement approach can be used in comparing different components of the service sector. Similarly, a review of a work by Pakdil and Aydin (2007) revealed the use of a modified version of SERVQUAL.

Abdullah, Jan, and Manaf (2012) utilized SERVPERF as a measurement approach in the evaluation of service quality perceived by airline passengers. In their study they found that the dimensions of tangibility, reliability, and assurance are most important.

Eboli and Mazulla (2007) have utilized original measurement methods in developing service measurement scale for bus transit. The same reveals three different quality dimensions covering service planning and reliability, comfort, ancillary factors, and network design. Comfort and ancillary factors appeared to influence the perceived satisfaction of passengers more than they do to other attributes.

Chou, Lou and Chang (2014) carried out an analysis involving high speed rail transit. They developed five dimension quality of service indicators using past results in transport studies. Indicators include perception of the tangible aspects of the cabin environment, convenience of the service, interactions with employees, and service reliability.

Lai and Chen (2011) have used an original measurement scale in investigating the intentions of individuals to make use of public transportation. They have revealed that service coverage, frequency, information and physical environment (i.e. cleanliness, safety and stability) influence the perceived satisfaction significantly.

Table 1:   An overview of research related to quality of transport sector

Author	Year	Mode	Approach	Quality of service dimension
Fick and Ritchie	*1991*	*Airline*	*SERVQUAL*	*Tangibles, reliability, responsiveness, assurance and empathy*
Pakdil and Aydin	*2007*	*Airline*	*SERVQUAL* *(modified)*	*Employees, tangibles, responsiveness, reliability and assurance, flight patterns, availability, image and empathy*
Abdullah, Jan and Manaf	*2012*	*Airline*	*SERVPERF*	*Tangibles, reliability, responsiveness, assurance and empathy*
Eboli and Mazulla	*2007*	*Bus*	*Original measure*	*Service planning and reliability, comfort and ancillary factors and network design*
Chou, Lou and Chang	*2014*	*Train*	*Original measure*	*Tangibles, convenience, employee interaction and reliability*
Lai and Chen	*2011*	*Public Transport*	*Original measure*	*Core services and physical environment*

2.5 BRT standard

BRT establishments are evaluated and certified using *BRT Standard* tool. The tool is based on international best practices and is a common point of global leaders in bus rapid transit design. It establishes a common definition of BRT and ensures that: *"BRT corridors more uniformly deliver world-class passenger experiences, significant economic benefits, and positive environmental impacts."* (ITDP, 2016) Overview of BRT Standard Scorecard and BRT Standard Rankings are summarized in this section:

2.5.1 OVERVIEW OF BRT STANDARD SCORE CARD

The BRT scoring system was put in place as a means of protecting the BRT brand and recognizing high-quality BRT corridors around the world. BRT corridors are certified as gold, silver, bronze or basic and thus set standards for the current best practices for BRT that are recognized internationally. Corridors are assessed in two ways: Design score and Full score (Design + Operations).

Design Score

It is *a basic reflection of quality of BRT corridor based on the implemented design and services.* The score indicates the maximum potential for performance on a corridor. The points are awarded for elements related to corridor design and those improving significantly BRT speed, capacity, reliability and quality of service.

Full score (Design + Operations)

These scores are most complete and offer realistic indicators of BRT corridor quality and performance. The score combines the Design score with operations deductions. In this, points are deducted from the score based on operational elements that significantly reduce corridor performance and service quality.

Point System criteria

The criteria used to determine the point system are as follows:

- *The points are proxies for better services (speed, capacity, reliability and comfort)*
- *The points are assigned based on consensus among BRT experts.*
- *The points should reward good, politically challenging design and operational decisions made.*
- *The metrics and weightings should be applicable easily and equitably.*
- *The basis of score should be transparent and independently verifiable.*

The maximum number of points a corridor can earn is 100.

2.5.2 BRT SCORECARD CRITERIA

The scorecard by ITDP (2012) showing criteria and point values that constitute the BRT standard is presented as figure 9 below. Each criterion is described in detail and point values attached.

CATEGORY	MAX SCORE		MAX SCORE
SERVICE PLANNING		**STATION DESIGN AND STATION-BUS INTERFACE**	
Off-board fare collection	7	Platform-level boarding	6
Multiple routes	4	Safe and comfortable stations	3
Peak frequency	4	Number of doors on bus	3
Off-peak frequency	3	Docking bays and sub-stops	2
Express, limited, and local services	3	Sliding doors in BRT stations	1
Control center	3		
Located in top ten corridors	2		
Hours of operations	2	**QUALITY OF SERVICE AND PASSENGER INFORMATION SYSTEMS**	
Multi-corridor network	2	Branding	3
		Passenger information	2
INFRASTRUCTURE			
Busway alignment	7		
Segregated right-of-way	7	**INTEGRATION AND ACCESS**	
Intersection treatments	6	Universal access	3
Passing lanes at stations	4	Integration with other public transport	3
Minimizing bus emissions	4	Pedestrian access	3
Stations set back from intersections	3	Secure bicycle parking	2
Center stations	3	Bicycle lanes	2
Pavement quality	2	Bicycle-sharing integration	1
		TOTAL	**100**

Figure 9: *BRT Standard scorecard. Source: The BRT Standard version 1.0 (ITDP, 2012)*

The weightings indicate the importance allocated to each criterion.

2.5.3 BRT STANDARDS RANKING

The four BRT standard points are categorised as basic, bronze, silver, and gold. A basic ranking means that the corridor meets the minimum criteria to qualify as BRT while bronze, silver, and gold reflect well-designed corridors that have achieved excellence.

Gold standard BRT (85 points or above)

Awarded to corridor that is consistent in almost all aspects of international best practices. Sufficient demand to justify BRT investment is pre-requisite.

Silver standard BRT (70-84.9)

This standard is achieved by a corridor that includes most of the elements of international best practices and is likely to be cost effective. There must be sufficient demand to justify BRT investment. Corridor must be achieving high operational performance and quality of service.

Bronze standard BRT (55-69.9)

This level is achieved by a corridor that solidly meets the definition of BRT and is mostly consistent with international best practices. A corridor with Bronze standards features some characteristics that elevate it above the BRT basics while achieving higher operational efficiencies of quality of service than that of BRT basic.

2.6 Urban transportation reforms in Tanzania

Urban transport in Tanzania has been provided by both the public and private sector. In the city of Dar es Salaam, public transport was provided by privately owned Dar es Salaam Motor Transport (DMT) between 1949 and 1967. Later, the DMT was succeeded by Usafiri Dar es Salaam (UDA) (SUMATRA, 2011). As the city kept on growing, UDA could not cope with the skyrocketing transport needs of Dar es Salaam dwellers (Nhundu. 2013) and the government in 1983 allowed private operators to offer services to bridge the gap between what UDA could provide and what is actually demanded by city transportations. In fact, UDA services continued to deteriorate towards 1986 giving prominence to private operators commonly known as "Daladala," which gladly filled in the void. Contrary to UDA buses built for special purposes of mass transportation and articulated buses in the early years,

Daladala operators used imported used buses with a passenger carrying capacity of not more than 40 (DART, 2014). The resulting phenomenon was poor quality of urban transportation services to Dar es Salaam city residents, increased traffic congestion, and increased air pollution.

Poor services of Daladala were also attributable to poor and unethical conduct of the drivers and bus conductors. These are mostly employed on daily basis and are given their salaries based on the number of trips they make per day. The bus owners set a minimum of collections per day that drivers have to hand to the bus owner; and drivers keep, as wages, whatever exceeds that required amount (SUMATRA, 2011). The non-synchronisation of ticketing amongst Daladala owners meant that if a bus breaks down then one has to board another bus and pay again. The existence of many individual operators who usually own a small fleet also poses a challenge to achieving a supply-demand balance of transport services in the city. The challenge is further compounded by operators with little understanding of relationships of key players in the industry (SUMATRA, 2011).

Moreover, the routes that provide access from one part of the city to another are not properly synchronised as the road networks and urban transportation licensing are under different jurisdictions. Licensing matters are under the Surface and Marine Transport Agency (SUMATRA) (SUMATRA, 2012) whereas roads fall under the Tanzania Roads Agency (TANROADS) and in other cases under respective municipalities. To address the shortfalls in urban transport and to improve living conditions for city residents, the Tanzanian government embarked on improving urban transportation by providing a reliable and efficient public transportation system. In this endeavour, the government opted for the Bus Rapid Transit (BRT) as the mode primarily because buses require low capital and are flexible as compared to the rail (Moylan *et al.*, 2013). Moreover, BRT systems have been operating successfully around the world (UITP 2010) including in some African cities such as Lagos and Johannesburg. The expectation was that the Dar es Salaam Rapid Transit (DART) would address all the negative aspects of the current services providers including matters related to ticketing, route synchronisation, and connectivity as well as quality of buses, safety, and good customer care.

2.7 Transport sector legislation

Institutional framework plays an important role in the state of public transport as a weak framework could lead to improper planning and, thereby, exacerbating congestion levels (Bruun, Del Mistro, Venter, & Mfinanga, 2015). In Tanzania, there are several pieces of legislation governing city transport (SUMATRA 2011). There is inadequate co-ordination amongst different government agents, which also explains why it has taken a long time for the DART to be realised (Bruun, *et al.*, 2015). There is also duplication in institutional arrangements (between ministries and authorities) resulting in gaps in the execution of responsibilities (JICA, 2008). Relevant legislation definitions or functions governing city transport are quoted as follows:

SUMATRA ACT of 2001

"This is an act that establishes regulatory authority in relation to surface and marine transport sectors, and to provide for its operation in place of former authorities and for related matters." The two sections relevant to this study are quoted as follows:

Section 6 (1) *(b) (ii) To establish standards for regulated goods and regulated services.*

Section 6 (b) (iv) *To regulate rates and charges.*

LOCAL GOVERNMENT ACT of 1982

"This is an act to make better provision for the establishment of urban authorities for the purpose of local government, to provide for the functions of those authorities and for other matters connected with or incidental to those authorities."

The following section of the act has an impact on urban transport,

Section 55 (1) (n).*To regulate the use and conduct of public vehicles plying for hire and their fares, to regulate the routes and parking places to be used by such vehicles, to appropriate particular routes, roads. Streets and parking places to specified classes of traffic, and when necessary to provide for identification of all licensed vehicles.*

ROAD TRAFFIC ACT of 1973

This act repealed and re-enacted the traffic ordinance. The section relevant to this study and influencing public transport is quoted as follows,

Section 73) a) *To regulate all traffic and keep order and prevent obstruction in all roads, parking places, thoroughfares or other places of public resort*

ROAD ACT of 2007

"This is an act to make provisions for road financing, development, maintenance, management, and other related matters." Sections related to this study are quoted as follows:

Section 4 (a) *To formulate road policy*

Section 4 (b) *To cause to be prepared and coordinate the implementation of road investment programmes.*

TRANSPORT LICENSING ACT of 1973

"This is an act to provide for transport licensing."

Its regulations establish that SUMATRA is responsible for putting in place procedures for issuance of licenses related to passengers and freight.

STANDARDS ACT of 2008

"This is an act for the promotion of the standardization of specifications of commodities and services, re-established the Tanzania Bureau of Standards and to provide better provisions for the functions, management, and control of the Bureau, to repeal the standards Act, Cap 130 and to provide for other related matters." A section relevant to this study is quoted as follows:

Section 4 (k) *Provide for the inspection, sampling and testing of locally manufactured and imported commodities with a view to determining whether the commodities comply with the provisions of this Act or any other law dealing with standards relevant to those commodities.*

DART AGENCY (Executive Agency Act of 1997)

A review of the Act establishes that the DART Agency has a principal task of modernising the urban transport and its facilities in the city of Dar es Salaam. DART is, therefore, tasked with the establishment and operations of BRT in Dar es Salaam. Moreover, DART is responsible for the orderly flow of BRT traffic in the city roads.

The review indicates that the provision of efficient public transport needs co-ordination and effective co-operation of all responsible legislative bodies as each has a role to play in the overall quality public service delivery. Since the policy objective of Dar es Salaam city is to promote public transport over private transport (Bruun *et al.*, 2015), paratransit systems should be part of an integrated system. This is because BRT alone is not a final solution. In addition, this is to ensure that conflicts in the urban transport sector are minimised when the existing operators are made part of BRT systems, a fact, which is supported by global practices (Arrivealive, 2017). In this regard, Dar es Salaam needs to integrate Daladala operations into the DART system so that Daladala system deliver effective services to feeder stations and, thus, improve the overall public transport. This means that the failure of the system would translate into the marginalisation and economic deprivation of the majority of the population (Afolabi 2016). Since Daladalas are regulated by SUMATRA (DART, 2014), there is a need for DART and SUMATRA to work together towards achieving the common goal of improving public transportation. To achieve the common goal, there is need of formulating effective policies that recognise the importance of the existing operators in making DART a success. The policies would engender the crucial complementarity to DART. Such complementarity would eventually be beneficial to many passengers, unlike the policies that tend to exclude the existing operators (Del Mistro & Behrens, 2015).

2.8 Quality of BRT services and user perception

The quality of service of BRT needs improvement to retain the passengers (particularly the choice passengers) as these would only continue to use the services when they perceive them as being of high quality (Hussein & Hapsari, 2014). As evidenced in Lagos, Nigeria, individuals tend to use their private vehicles due to lack of satisfaction with their public transport services (Afolabi, 2016). In this regard, the literature reviewed has not provided evidence of the assessment of current BRT services in Tanzania. There is limited research work on the general quality of public transport in Dar es Salaam; however, there is evidence that studies on the assessments of the quality of BRT services and user perception have been carried out in other cities where BRT services are provided. Attributes of quality of BRT services are discussed by considering convenience, efficiency, passengers' comfort, safety, security, reliability, and passenger care.

2.8.1 AFFORDABILITY

In modelling, commuter preference for the then proposed bus rapid transit in Dar es Salaam; Nkurunziza *et al.* (2012b) establish that lower fares attract passengers leading to sustainability of the public transport services. Nhundu (2013), who reviewed Zeithaml's *et al.* (2006) study, observes that affordability is related to the degree to which a customer believes that the service is affordable. Affordability and decision-making by customers are related. In this regard, Okagbue, Adamu, Iyase, and Owoloko (2015) have recommended for a review of fares as part of the means for ameliorating the challenges faced by commuters.

2.8.2 CONVENIENCE

Xiyuan (2014) analysed the service quality of BRT by presenting a case study of Hefei. The study indicates that convenience is the most important aspect of the BRT service. Nhundu (2013) defines convenience as the level of easiness by which a customer can use a particular service. Convenience is related to both the level of quality of the service coupled with the ease-of-use. Thus, convenience is attributable to affordability, hours of service, frequency of service, waiting time, service coverage, travel speed travel time on buses, and ticketing.

Hours of service

Longer service hours throughout the day or a week elevate the level of service of a viable transit corridor (ITDP, 2016). Hours of service has an impact on passengers' convenience, as passengers plan their movements according to the times when buses are available. While studying the introduction of BRT in Asian developing cities, Satiennam, Fukuda, and Oshima (2006) found long hours of coverage as an important characteristic of the BRT services. For BRT to give an impression that the service is true rapid transit, hours of service should be 18 to 20 hours. This time would ensure service availability nearly most of the times of the day and, thus, meet the transportation demand (APTA 2010).

Frequency of service and waiting time

Satiennam *et al.* (2006) revealed that a high service level is attributable to shorter headways, that is, of high service frequency, which allows customers to make decisions pertaining to using the system. In fact, waiting time for the service at the terminals has an impact on the customers' journey planning. In this regard, Amiegbebhor and Dickson (2014) indicate that waiting time is an area demanding attention when considering customer satisfaction. Lai and Chen (2011) cited frequency as a factor influencing significantly on passengers' perception of the service.

Service coverage

In their study on challenges facing commuters, Okgabue *et al.* (2015) recommend an increase in the number of routes to ease the headaches commuters faced in Lagos, Nigeria. In a similar vein, Nhundu's (2013) review of Zeithaml *et al.* (2006) indicates that the ease-of-access to service reflects the customers' ability to get to the service provider and get the service conveniently and quickly. Lai and Chen (2011) have found service coverage to have significant effect on the perception of quality of service.

Ticketing

Reviewing service quality dimension of commuter uptake in Cape Town, South Africa, Ugo (2014) found that information on where to purchase cards is an item of dissatisfaction highlighted amongst commuters. Difficulties in purchasing or reloading cards may lead to loss of passengers. A convenient BRT system would thus be the one in which purchasing tickets/cards is not difficult.

2.8.3 EFFICIENCY

In Nhundu's (2013) review of Zeithaml *et al* (2006), efficiency is defined as how simple the service is, and how the service is structured. Inefficiency in service delivery has negative effects on the quality of the service (Nhundu, 2013).

Travel time on buses

In modelling commuter preference, for the then proposed bus rapid transit in Dar es Salaam, scholars (e.g. Nkurunziza, Zuidgesst, Brussel & Maarseveen, 2012b) established that the sustainability of transport services and its attractiveness to the public depend on the shortened travel time. This travel time is the duration passengers take before reaching their respective destinations.

Travel speed

Travel speed affects both the time it takes for passengers to reach their respective destinations and the passengers' safety (discussed under item 2.8.6). As established by Mahmoud, Verdinejad, Jandaghi and Mughari (2010), there is a significant relationship between BRT speed and customer satisfaction.

Branding

Branding can influence passengers' decision on choosing from many alternatives. This is particularly because customers need alternative ways of achieving or accomplishing what they want (Ugo, 2014). In the assessment of public perception of the BRT in San Francisco, California in the United States Moylan *et al.* (2013) suggested that the success of a system is related to how the services are provided and marketed. In this regard, effective branding assists in marketing (Entrepreneur, 2017). BRT is generally a premium investment, which ought to be differentiated from traditional transport systems through unique branding (APTA 2010). The Institute for Transportation and Development Policy (ITDP) (2016) considers strong branding as an important aspect of quality of service.

Information

Information on various aspects of the BRT system is important as it informs passengers about the service available. Lai and Chen (2011) cite information as an aspect that influence passenger's perception on service quality. In this regard, Ugo (2014) observes that passengers need to be well informed about a service or a product specifically with regard to the attendant policies and procedures. In another study, Sivakumar, Yabe, Okamura, and Nakamura (2006) found that understanding how to use the BRT system was a challenge to residents in developing cities. Moreover, information displays are necessary for efficient services (Eboli & Mazulla, 2007). The facts available on the displays underscore the importance of providing information to the users. Useful information includes system maps, which assist customers in planning their journeys. Furthermore, directions to get to various BRT facilities, including feeder stations are also part of information displayed. Thus, providing customers with relevant information is critical to their overall positive experience of high quality service (ITDP, 2016).

2.8.4 PUNCTUALITY

Patel and Makwana (2015) in their review of the assessment by Purcher, Park and Kim (2005) indicate that customer satisfaction is linked to timeliness, accuracy and completeness. Reliable service delivery at the right time is the least the BRT riders demand.

According to Nhundu (2013) in the review of Zeithaml's *et al.* (2006) work, reliability of the service is related to correct technical functioning. .Moylan *et al.* (2013) also found that achieving reliability of BRT services requires an appropriate combination of technical features such as level boarding, signal priority, dedicated lanes and off-board ticketing.

Technical functioning

Technical functioning relates to the condition of the buses in use in a given BRT system. Buses that are well serviced are not expected to breakdown en route. If passengers frequently fail to reach their intended destinations because buses keep on breaking down, then the service would be regarded as unreliable. .

2.8.5 PASSENGER COMFORT

Passenger comfort is related to amenities the BRT provides as part of its infrastructure. During modelling of the commuter preference for the then proposed bus rapid transit in Dar es Salaam, Nkurunziza *et al.* (2012b) established that sustainability of transport services and their attractiveness to the public depends on them providing the much needed comfort which is often absent from the conventional public bus transport systems. Amenities such as shelters, benches, refreshment stalls, and toilets constitute elements of comfort that are necessary to ensure that passengers satiate their needs at the terminals while waiting for services. Comfort is also related to the ergonomics of the buses in use. In this regard,

Mahmoud *et al.* (2010) have pointed out the interrelatedness of ergonomics and customer satisfaction.

Shelter

Adequate shelter is necessary so that passengers get protection from adverse conditions such as the sun and rain. Eboli and Mazulla (2007) also used shelter as a factor in determining customer satisfaction. Stations need to be protected from weather (wind, rain, heat or snow as appropriate) (ITDP, 2016).

Benches

Although it is not possible and practical to provide seating facilities to all waiting passengers, a good number of benches are necessary to cater for children, the sick, the elderly and the infirm while waiting for the BRT service. Eboli and Mazulla (2007) also used benches as a factor in determining customer satisfaction.

Interior of buses, Seating,

The interior of buses related to seating, entering and exiting buses has an impact on customers' convenience. Ziyari and Hajisharifi (2013) suggest that removing obstacles such as poor interior quality of BRT buses could enhance attractiveness of the service from a customer's point-of-view

2.8.6 SAFETY AND SECURITY

While evaluating passenger comfort in BRT systems, Batarce, Munoz, Ortuzar, Raveau, Mojica and Rios (2015) established that safety and security are the most important attributes of the travel mode. Furthermore, ITDP (2016) lists safe and comfortable station environment as an important aspect of high-quality BRT service. Safety is related to accidents (What are the chances that buses would be involved in accidents?) whereas security is related to crime (What are the chances that passengers would be attacked, robbed, sexually harassed, etc.?).

Safety on bus

The safety of the transport system influences the safety of society (Ugo, 2014). In fact, safety on a bus is a key to the overall safety of a transport system. This aspect has also been highlighted by Patel and Makwana (2015) who stressed that safety is one of the factors demanding consideration for further improvement of BRT services. Lai and Chen (2011) consider safety as a factor influencing the perception of service quality by passengers.

Security at stations/terminals

In urban areas, security is always a concern before customers decide upon using a particular facility. In this regard, Okagbue *et al.* (2015) call for adequate security at BRT bus stops in Lagos to make the service more attractive to city residents. Well-lit and secured stations (by guards or cameras) can influence maintaining ridership (ITDP, 2016).

2.8.7 PASSENGER CARE

Behaviour of driver and other crew

Mahmoud *et al.* (2010) have established that the drivers' behaviour influence passenger satisfaction. Similarly, Okagbue *et al.* (2015) assert that the attitude and behaviour of drivers affect passenger satisfaction. Moreover, Ugo (2014) indicates that friendliness is an important basic aspect related to politeness and courtesy during greetings.

SUMATRA (2011) established that the majority of the passengers in the city of Dar es Salaam were not satisfied with city transport services due to poor conduct of bus crews and mainly conductors. Most notable is the abusive language that bus conductors' use poor hygiene of the buses, dirtiness of crew uniforms, overloading during peak hours and high turned-on volumes of music and radio all of which can put off some riders.

Effects of Previous transactions

Customer satisfaction is strongly related to previous transactions (Klein, 2014). In other words, passengers would willingly return for the service if they were satisfied with the previous service. For a particular service operator such as DART it is, therefore, important to keep on improving on all the quality attributes discussed as part of concerted efforts aimed at retaining passengers and, specifically, in the corridors where alternatives do exist.

2.8.8 SUMMARY

The attributes discussed above contribute to how passengers perceive their satisfaction with respect to BRT services. A summary of the factors and sub-factors is presented in Table 2:

Table 2: Summary of factors and sub-factors with their frequency in the literature reviewed

Service factor	Service sub- factor	References
Affordability		Nkurunzia et al. (2012), Okgabue et al(2015)
Convenience	Hours of service	Patel & Makwana (2015)
	Frequency of service	Patel & Makwana (2015), Lai & Chen (2011)
	Waiting time	Amiegbebhor & Dickson (2015)
	Service coverage	Okgabue et al. (2015), Lai & Chen (2011)
	Ticketing	Ugo (2014)
Efficiency	Travel speed	Patel & Makwana (2015)
	Travel time on buses	Nkurunzia et al. (2012)
	Branding	Ugo (2014), Bartace et al.(2015), Lai & Chen (2011), ITDP (2016)
	Availability of information	Ugo (2014), ITDP (2016)
Comfort	Shelter	Eboli & Mazulla (2007), (ITDP (2016)
	Benches	Eboli & Mazulla (2007)
	Interior of buses	Ziyari & Hajisharrifi (2013)
	Seating	Okgabue et al. (2015)Eboli & Mazulla (2007)
	Entering and exiting	Eboli & Mazulla (2007)
Safety and security	Safety on bus	Ugo (2014), Bartace et al.(2015), Lai & Chen (2011)
	Security at stations	Okgabue et al. (2015), Bartace et al.(2015), ITDP (2016)
Punctuality	Technical functioning	Patel & Makwana (2015), Nhundu (2013)
	Accuracy	Nhundu (2013)
Passengers care	Drivers/crews behaviour	Mahmoud et al. (2010), Ugo (2014), SUMATRA (2011)
	Effects on previous transactions	Klein (2014)

The summary of factors in Table 2 served as a guide in questionnaire formulation. These are tabulated and presented in Table 3:

Table 3: Motivation behind Factors Selected for Questionnaire Formulation

Factors	Motivation
Drivers' and crew behaviour	This sub-factor will be included in the questionnaire even though it is not considered as the most important sub-factor considering that passengers have no transport choice alternatives in the areas where DART operates.
Affordability Seating Alighting precision	These sub-factors have appeared with considerable frequency in the literature reviewed.
Frequency of service Service coverage Travel Speed Entering and exiting buses Reliability Security at stations Shelter	These factors are necessary as they characterise DART service.
Hours of Service Ticketing	These sub-factors have appeared with less frequency but they are, nevertheless, key to service access.

The motivation behind the factors facilitated the development of a questionnaire deployed in this study (see Appendix I). As Tanzania has two major languages operating at the national level—Kiswahili as a national language and English as an official language and language of instruction in the country's higher level of education, the questionnaire was translated into Kiswahili to cater for non-English speaking passengers.

2.9 Analytical methods

Some common analytical methods in the literature are discussed hereunder.

2.9.1 STATED PREFERENCE

The Stated Preference (SP) method was developed in marketing research to handle hypothetical situations before it was adopted in the transportation field (Sivikumar *et al.*, 2006). . In the preparation of the survey design to grasp and compare the users' attitudes to the bus rapid transit in developing countries, Sivakumar*et al.* (2006) used this method alongside the Revealed Preference (discussed as the next item). They used the method as their study was mainly guided by hypothetical situation. SP was also used by Nkurunziza *et al.* (2012b) while modelling commuter preferences for the then proposed BRT in Dar es Salaam. Nkurunziza *et al.* (2012a) deployed SP while studying the spatial variation of the Transit Service Quality Preference in Dar es Salaam.

2.9.2 REVEALED PREFERENCE

Revealed preference (RP) is the analysis method that entails observing the existing situation (actual market) (Sivakumar *et al.*, 2006). Researchers employed this method alongside SP. They used a few RP questions for completeness and cross checking. Bartace *et al.* (2015) used RP while evaluating the passengers' comfort in the BRT system.

2.9.3 GARRETT'S RANKING TECHNIQUE

According to Dhanavandan (2016), Garrett's ranking technique is a method where all factors under a particular study are assigned ranks by the respondents. The outcome of ranking is converted into score value with the help of a formula:

Per cent position = $100(R_{ij}-0.5)/N_j$

Where:

R_{ij} = Rank given for the ith variable by jth respondent

N_j = Number of variable ranked by jth respondents

The estimated per cent positions are converted into scores with the help of Garrett's table. The scores for each individual are added for each factor. The total value of scores and mean scores are calculated. The factor with the highest mean score is recorded as the most important.

This method is applicable when looking for the most significant attributes or factors influencing the respondents (Mehta 2011).

2.9.4 LIKERT SCALES

Likert Scale is a five (or seven) point scale which allows the respondents to state how much they agree or disagree with a particular statement. *Bertram (2007) defines Likert scale as "a psychometric response scale primarily used in questionnaires to obtain participants preference or degree of agreement with a statement or set of statements."* Every response is assigned a numerical score, which indicates how favourable or unfavourable it is; the scores are then totalled to measure the respondent's attitude. Ugo (2014) used the Likert scale in his study entitled *"The bus rapid transit system: A service quality dimension of commuter uptake in Cape Town, South Africa"*. Okgabue *et al.* (2015) also employed this analytical method in their study *"Motivations and challenges faced by commuters using bus rapid transit in Lagos, Nigeria."* The method has also been used by Moylan *et al.* (2013) in their study on public perceptions of the bus rapid transit in the San Francisco Bay area in the US.

2.9.5 PREFERRED ANALYTICAL METHOD

As this study is on measuring user perception as determined by several factors, and considering the formulation of the questionnaire, as appended, the Likert's scale analytical method was adopted. The method was used principally because:

i) This study has concentrated on satisfaction.

ii) It is considered more reliable as the respondents are expected to answer each statement included in the questionnaire (Mehta, 2011)

iii) It can be easily used in the respondent-centered studies.

iv) It is frequently used in opinion research (Mehta, 2011)

For the purpose of this study a five point scale was used in which .1 is most positive and 5 is least positive.

CHAPTER THREE

3.0 METHODOLOGY

3.1 Study Area

Dar es Salaam city constitutes the study area because of the existence of DART services. Specifically, the study focused on Ilala, Kinondoni, and Ubungo districts where the DART was currently operating. . These areas have been covered by DART under the first phase. Among the five districts found in Dar es Salaam region Temeke and Kigamboni are yet to be covered by DART.

3.2 Study population

The study consisted of DART users who have paid for the bus services.

3.3 Sample size determination

The sample size was determined using the following formula.

Necessary Sample Size = (Z-score)^2*Proportion sample "positive"*(1-Proportion sample "positive")/(margin of error)^2

Where;

Z-score is a constant value and assumed as 1.96 corresponding to 95 per cent confidence level.

Proportion sample is assumed as 0.5 for 50 per cent

Margin of error (confidence interval) is assumed as 0.06 for +/- 6.0 per cent

The sample size is calculated as follows:

Sample Size = $(1.96)^2$ x $(0.5)^2$ x $/(0.06)^2$ = 266.7.

About 267 respondents were needed. The study drafted in 260 respondents. This results in a margin of error of 0.61.

3.4 Sampling procedure

3.4.1 SELECTION OF STUDY SITES

Study sites such as Terminals, Trunk stations and on board buses were purposively selected due to the ease of identifying DART passengers. Stations were randomly selected. Off-peak hours were chosen for interviews inside the buses. Based on personal experience, it was difficult to carry out such interviews during peak hours when the buses were characterised by overcrowding. In addition, early morning and evening hours (usually peak) were chosen for interviewing passengers as they waited for buses at the terminals (Kivukoni, Kariakoo, Morocco, Ubungo, and Kimara (Figure 5). From personal experience, it was apparent that passengers experienced a waiting time of up to 30 minutes before the buses arrived at the terminals during peak hours. Thus, the researcher seized this as an opportunity for carrying out interviews relatively as the potential interviewees could spare a moment while waiting for their respective buses.

3.4.2 SELECTION OF THE STUDY POPULATION

Convenience sampling method was used to select the respondents for structured interviews. Etikan, Musaand Alkassim (2016) define convenience sampling as *"a type of non-probability or non-random sampling where members of target population that meet certain practical criteria, such as easy accessibility, geographical proximity, availability at a given time or the willingness to participate are included for the purpose of the study"*. DART users waiting or leaving the stations were approached for interviews. Users on board buses were also approached for interviews during the non-peak hours. A structured interview was chosen since it does not require advanced skills from interviewers and can be accomplished at relatively lower costs (Kothari, 2009).

3.5 Data collection Procedures

Data were collected daily from 6.00am to 6.00pm continuously for a period of two weeks (all days including weekends). The data collection instrument was a semi-structured interview guide with both open and closed-ended questions (Appendix I-English version, Appendix II-Kiswahili version). The author collected data with the help of four research assistants who were selected because they had attained a minimum of secondary ordinary education and were fluent in both (written and spoken) Kiswahili and English. Research assistants were briefed on the objectives of the study and were properly trained on how to communicate with the respondents before embarking on data collection. The interviewers visited DART stations and terminals and interviewed the respondents as they waited for their buses. They asked questions and filled out the interview forms themselves. Interviewers also boarded the buses on different routes and interviewed users on board as they travelled from one destination to another. This approach was considered as more practical than simply giving the respondents forms and collecting them upon completion which appears to pose a challenge as most users rush to get out of buses when they reach their destination and may not have taken time to hand back the forms. The interview questionnaire was piloted and improved upon based on the results of piloting before it was administered in field.

3.6 Data Management analysis

The completed questionnaires were then checked every evening by the researcher to determine whether they had been filled out correctly and completely. The data obtained were then coded manually using Microsoft Excel software. The researcher validated and analysed data using *Stata version 13.0* software.

3.7 Ethical clearance

Ethical clearance was sought and obtained from EBE EiR committee as required by the Faculty of Engineering and the Built Environment (EBE) before embarking on research work (Appendix III). Formal permission to conduct the study within the BRT infrastructure and buses was obtained from DART management (Appendix IV) as they were satisfied that the researcher is a bona fide student of the University of Cape Town (UCT). Informed consent was obtained from the respondents who were briefed beforehand about the purpose of the study. A copy of the informed consent form is included as Appendix V.

CHAPTER FOUR

4.0 RESULTS

4.1. General findings

4.1.1 SOCIO-DEMOGRAPHIC CHARACTERISTICS

Overall, 260 participants took park in the study. Their socio-demographic characteristics are shown hereunder:

Table 4: Socio-demographic characteristics of the study participants

Question number	Characteristics	N	(%)
25 (i)	**Age group**		
	18-25	76	(29.2)
	26-35	93	(35.8)
	36-45	69	(26.5)
	45+	22	(8.5)
25 (ii)	**Sex**		
	Male	128	(49.2)
	Female	132	(50.8)

The composition of the sampled users (Table 4 and Figure 10) shows that 49.2 per cent were males and 50.8 per cent were females. The age structure reveals that 65 per cent of the DART users who took part in the study were aged between 18 and 35 years (Figure 11).

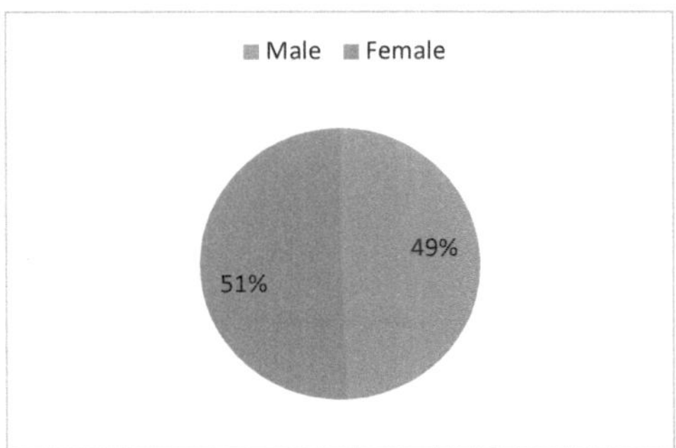

Figure 10: Gender distribution among DART passengers interviewed

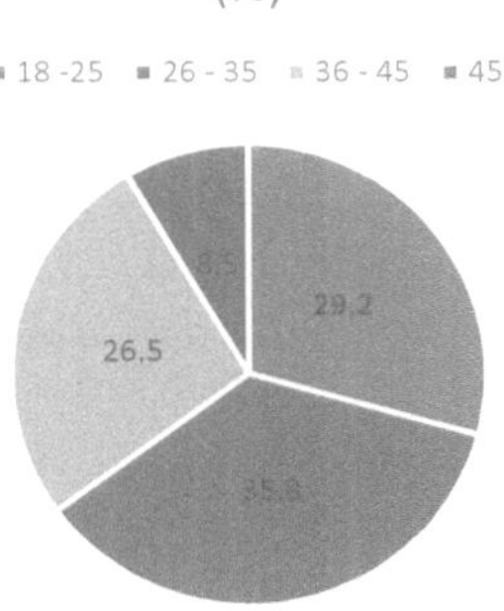

Figure 11: Age group distribution among DART passengers interviewed

4.1.2 GENERAL PERCEPTION OF DART SERVICES

The general perception of DART services is positive as the results presented in Table 5 and Figure 12 illustrate that 61.0 per cent of passengers are satisfied and 16.0 per cent find the services as being of average quality. 23.1 per cent of passengers were not satisfied with the services provided. Similar findings are reported by Deng and Nelson (2012) in their passenger survey from Beijing Southern Axis BRT Line1 where captive passengers indicated overall satisfaction with BRT services.

Table 5: General passengers' perception of DART services

Overall experience of using DART services (Question 23)	N	(%)
Very satisfactory	2	0.8
Satisfactory	154	60.2
Average	41	16.0
Unsatisfactory	57	22.3
Very unsatisfactory	2	0.8

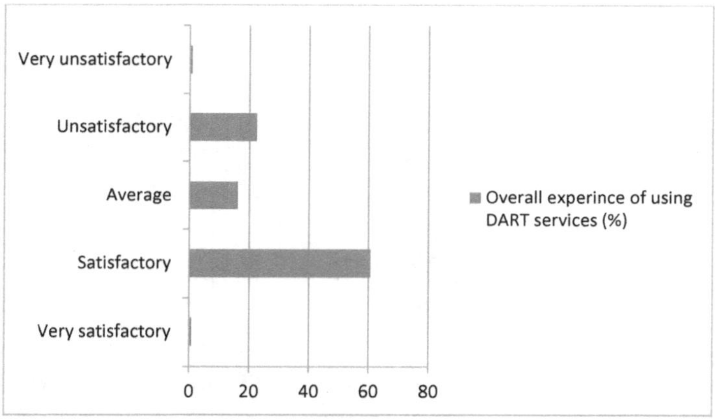

Figure 12: General passengers' perception of DART services

4.2 Participants' perception of the value of service quality attributes

A rating scale of 1 -5 was applied in the survey in which 1 was the most positive and 5 the least positive response (except for questions 8 and 22). The results of passengers' perception on the value of the existing service quality attributes are presented in Table 6. (The Analysis of the attributes with respect to means and standard deviations is presented in Table 7 and in Figure 15). .

Overall, different perception statements have been used in different questions as shown in the questionnaire used (Appendix I). The word "Very" is interpreted as positive or negative depending on how a question was asked. Where a negative word follows (e.g. "inconvenient" [Q5], "inadequate" [Q10]) it is interpreted as a negative score.

The response words used for middle option are "Average," "Few times," "Not sure" and "Once in a while." In order to determine the positive or negative score of a particular service attribute, the middle words have been interpreted as follows:

a). "Average" as positive.

b). "Few times" as positive if the more times would be detrimental and negative if more times would be beneficial

c). "Not sure" as negative OR NEUTRAL

d). and "Once in a while", as positive if more times would be positive if less times would be detrimental and negative

The colour codes below have been applied to Table 6 for the interpretation of the attributes.

Colour	**Interpretation**
	Very Positive
	Positive
	Neutral
	Negative
	Very Negative

Table 6: Participants' perception of the value of service quality attributes

Question number	Attributes	N	(%)
1	**Affordability of fares**		
	Very affordable	12	(4.6)
	Affordable	84	(32.3)
	Average	93	(35.8)
	Not affordable	43	(16.5)
	Highly Unaffordable	28	(10.8)
2	**Comfort inside the buses**		
	Very comfortable	2	(0.8)
	Comfortable	65	(25.0)
	Average	50	(19.2)
	Not comfortable	117	(45.0)
	Highly Uncomfortable	26	(10.0)
3	**Frequency of arriving at the destination within planned time**		
	Always	12	(4.6)
	Most of times	100	(38.6)
	Few times	99	(38.2)
	Very few times	45	(17.4)
	Never	3	(1.2)
4	**Frequency in relation to travel needs**		
	Highly satisfactory	3	(1.2)
	Satisfactory	162	(62.3)
	Not sure	45	(17.3)
	Unsatisfactory	49	(18.8)
	Highly unsatisfactory	1	(0.4)
5	**Convenience of the service hours**		
	Very convenient	17	(6.7)
	Convenient	178	(70.7)
	Not sure	26	(10.3)
	Inconvenient	31	(12.3)
	Highly inconvenient	0	(0.0)
6	**Time to reach destinations**		
	Very short	11	(4.3)

Question number	Attributes	N	(%)
	Short	111	(43.2)
	Average	105	(40.8)
	Long	28	(10.9)
	Very Long	2	(0.8)
7	**Is it safe to enter/exit the buses**		
	Yes	210	(81.7)
	No	47	(18.3)
8	**Buses' breakdown**		
	Very often	0	(0)
	Often	23	(8.8)
	Not sure	57	(21.9)
	Less often	57	(21.9)
	Very less often	122	(46.9)
9	**Adequacy of security at DART stations**		
	Very Adequate	8	(3.1)
	Adequate	78	(30.5)
	Average	148	(57.8)
	Inadequate	20	(7.8)
	Very inadequate	2	(0.8)
10	**Adequacy of shelter at DART stations**		
	Very Adequate	7	(2.7)
	Adequate	121	(46.9)
	Average	98	(38.0)
	Inadequate	31	(12.0)
	Very inadequate	1	(0.4)
11	**Is it easy to buy a DART ticket**		
	Yes	166	(64.6)
	No	91	(35.4)
13	**Does BRT infrastructure provide universal access?**		
	Yes	206	(81.1)
	No	48	(18.9)
17	**Frequency of service usage**		
	Very often	28	(11.0)
	Often	159	(62.4)
	Once in a while	36	(14.1)

Question number	Attributes	N	(%)
	Less often	24	(9.4)
	Very less often	8	(3.1)
18	**Do you own a private car?**		
	Yes	50	(19.4)
	No	208	(80.6)
Question number	**Attributes**	**N**	**(%)**
19	**Do you see yourself continuing using DART services?**		
	Yes	249	(96.9)
	No	8	(3.2)
22	**While at the bus stop, how often does the bus pass and you cannot get on board?**		
	Very often	10	(4.0)
	Often	45	(17.9)
	Once in a while	111	(44.0)
	Less often	63	(25.0)
	Very less often	23	(9.1)

Findings presented in Table 6 are discussed as follows:

4.2.1 CONTINUED USE OF DART SERVICES (Q19)

96.9 per cent of the passengers interviewed consider continuing using DART services. Since 80.6 per cent of the passengers interviewed are captive riders, some choice riders also consider continuing using DART services.

4.2.2 ADEQUACY OF SECURITY AT STATIONS (Q9)

The study analysis indicates that only 8.6 per cent of the respondents found security at stations as inadequate whereas 33.6 per cent reported it to be adequate. About 57.8 per cent reported security to be average. "Average" has been interpreted as a positive perception bringing the total positive perception to 91.4 per cent

4.2.3 TIME TO REACH DESTINATION (Q6)

The study analysis indicates that the majority of the respondents found time to reach their destination as acceptable (88.3%). Only 11.7 per cent found the time as long. "Average" has been interpreted as a positive perception.

4.2.4 ADEQUACY OF SHELTER AT STATIONS (Q10)

The study analysis reveals that only 12.4 per cent of the respondents' found the shelter at stations to be inadequate, 2.7 per cent found it very inadequate. Another 49.6 per cent found security at the shelters as adequate while 38.0 per cent found shelter as average. "Average" has been interpreted as a positive perception bringing the total positive perception to 87.6 per cent.

4.2.5. SAFETY WHILE ENTERING OR EXITING BUSES (Q7)

On this variable, 81.7 per cent of the passengers interviewed reported that it was safe to enter or exit the buses.

4.2.6 ACCESSIBILITY OF BRT INFRASTRUCTURE (Q13)

With regard to accessibility, 81.1 per cent of the respondents indicated that BRT infrastructure provided universal access. Implicitly, consideration was largely taken for the physically challenged passengers as well.

4.2.7 FAILURE TO GET ABOARD BUSES (Q22)

With regard to failing to get on board the DART buses, 34.1 per cent of the passengers interviewed reported to have gotten on board the buses often whereas 21.9 per cent reported to have failed to do so often and 44.0 per cent fail to get inside the buses once in a while as buses pass the stations. "Once in a while" has been interpreted as positive bringing the total positive perception to 78.1 per cent.

4.2.8 CONVENIENCE OF SERVICE HOURS (Q5)

The study analysis shows that more than half (77.4%) of the passengers interviewed found the DART service hours as convenient whereas another 12.3 per cent found them as inconvenient. 10.3 per cent were not sure whether the service hours were either convenient or not to them

4.2.9 FREQUENCY OF SERVICE USAGE (Q17)

About 73.4 per cent of the DART users reported to be frequent riders. 13.8 per cent use services once in a while and 12.2 per cent of the interviewed users indicated using services less often. "Once in a while" has been interpreted as negative.

4.2.10 AFFORDABILITY OF FARES (Q1)

The study analysis shows that 72.7 per cent of the respondents found the fares as affordable (as only 27.3 per cent found the fares as unaffordable). Therefore, the fares charged are considered as acceptable by the majority of DART passengers. "Average" has been interpreted as a positive perception.

4.2.11 BUSES' BREAKDOWN (Q8)

About 68.8 per cent of the respondents reported to have had witnessed the buses breaking down "less often" (21.9 per cent-less often, 46.9 per cent-"very less often"). About 21.9 per cent of riders are not sure whether they have seen buses breakdown or not. The total positive perception for this attribute is 68.8 per cent.

4.2.12 EASINESS OF BUYING DART TICKETS (Q11)

Some 64.6 per cent of the respondents indicated that it was easy to buy DART tickets whereas 35.4 per cent reported that it was difficult. Similar results were reported by Amiegbebhor and Dickson (2014) while assessing impact of BRT on commuter satisfaction in Lagos, Nigeria, where ticket purchasing was not a problem. In case of DART, long queues and few ticket attendants were cited as the reason for the reported difficulty. Long queues are common in the main DART terminals, especially during peak hours.

4.2.13 FREQUENCY IN RELATION TO TRAVEL NEEDS (Q4)

The study analysis also reveals that 63.5 per cent of the respondents were satisfied with the frequency of DART services in relation to their travel needs. On the other hand, 19.2 per cent were not satisfied with frequency in relation to their travel needs and 17.3 per cent were not sure of the relationship of frequency and their travel needs.

4.2.14 COMFORT INSIDE THE BUSES (Q2)

Study analysis indicates that 55 per cent of the respondents reported that the inside of the buses as not comfortable. "Average" has been interpreted as a positive bringing positive perception to 45 per cent.

4.2.15 ARRIVING WITHIN PLANNED TIME (Q3)

The study analysis shows that only 43.2 per cent of the respondents found the arrival time to be within the acceptable limits. Some 38.2 per cent arrived within the planned time few times, 17.4 per cent arrived within the planned time very few times whereas 1.2 per cent had never arrived within the planned time. "Few times" (17.4%) has been interpreted as a negative perception.

4.2.16 CARS OWNERSHIP AMONGST DART PASSENGERS (Q18)

With regard to car ownership, 19.4 per cent of passengers interviewed owned cars. This means that nearly a quarter of the passengers are choice riders. With increased comfort and frequency, more choice riders may be attracted to the service. If this happens DART would be contributing significantly to the reduction traffic congestion in the city of Dar es Salaam.

4.3 Means of participants' perceptions on service quality attributes

Standard deviations and means of service quality attributes have been presented in Table 7. This is followed with, a discussion with a sequence from the most positive to the most negative response with adjustments being made for questions 8 and 22.

Table 7: Mean distribution of service quality attributes

Question	Attributes	Standard deviation	Mean
1	Affordability of fares	1.05	2.58
2	Comfort inside the buses	0.99	3.29
3	Arriving within planned time	0.84	2.72
4	Frequency in relation to needs	0.82	2.55
5	Convenience of Service hours	0.77	2.35
6	Time to reach destination	0.77	2.61
8	Buses breakdown	1.02	*4.32 translates to 1.68
9	Adequacy of Security at stations	0.68	2.71
10	Adequacy of Shelter	0.75	2.64
17	Frequency of use of services	0.90	2.39
22	Failure to get on board as buses pass	0.96	*3.12 translates to 2.88

*A positive score for Questions 8 and 22 is 5 whereas it is 1 for all other questions.

Specific analyses by age and gender are presented in Tables 8 and 9.

For the scores in Table 7, all attributes have a positive perception except "comfort inside buses" which has a negative perception. (Actual scores for questions 8 and 22 were 4.32 and 3.12 respectively).

4.3.1 BUSES BREAKING DOWN (Q8)

This attribute has the highest level of satisfaction (inverted mean score 1.68, actual score is 4.32). This means that break down of buses is witnessed very few times by the passengers interviewed. This result could be attributable to the operator using the buses, which are relatively new. However, to keep up with the result observed (i.e. less frequency of buses' breakdown) it would be necessary for the operator to devise a comprehensive and adequate maintenance plan and adhere to it.

4.3.2 CONVENIENCE OF SERVICE HOURS (Q5)

This attribute has a mean score of 2.35, hence indicating that the passengers interviewed found the service hours as convenient. In other studies such as one on future preferences and satisfaction of BRT (see Patel & Makwana, 2015) customers in Ahmedabad City recommended a 24-hour service. This recommendation could also assist DART while rolling out further phases so that a higher percentage of users can be satisfied with the service hour.

4.3.3 FREQUENCY OF USE OF SERVICES (Q17)

This attribute has a mean score of 2.39, indicating that the majority of passengers interviewed use the services frequently as indicated by 71.9 per cent of the passengers interviewed. DART has, therefore, managed to keep more than two-thirds of its passengers.

4.3.4 FREQUENCY IN RELATION TO NEEDS (Q4)

This attribute has a mean score of 2.55, indicating a positive response. However, efforts should be made to increase the frequency so that a higher mean score is achieved. In this regard, DART should strive to reach a higher satisfaction level to begin to match with that of Cape Town's MyCiti (Ugo, 2014). Increasing the number of buses would increase frequency and help ease congestion.

4.3.5 AFFORDABILITY OF FARES (Q1)

This attribute has a mean score of 2.58, indicating that the majority of passengers interviewed perceive the fares charged as average. However, the score is at median and, thus, does not represent a positive or negative perception. Efforts should be made so that a higher positive score is recorded to indicate higher acceptance of the fares charged.

4.3.6 TIME TO REACH DESTINATION (Q6)

This attribute has a mean score of 2.61, which is above average but not as high as 1.68 score (inverted, actual score 4.32) for Q8. With very few bus breakdowns observed (see sub-section 4.3.1), there are other factors contributing to lowering BRT rider satisfaction. A personal observation reveals that although DART buses operate on dedicated lanes the traffic signals are sometimes (especially during peak hours) interrupted by traffic police officers who do personal judgement to regulate the overall traffic flow. If such DART buses were not interrupted, more passengers would be satisfied with the time it takes them to reach their respective destinations.

4.3.7 ADEQUACY OF SHELTER (Q10)

This attribute has a mean score of 2.64, indicating a positive score though not very high. Shelter should be improved so that riders can perceive this attribute positively. Shelter can be boosted by simply improving sitting facilities inside the DART stations. A personal experience inside DART stations found enough space for accommodating more sitting facilities (benches). Terminals appear to be provided with benches while Trunk stations do not have any benches.

4.3.8 ADEQUACY OF SECURITY (Q9)

This attribute has a mean score of 2.71, indicating a positive score though not very high. Security should be improved so that a much lower percentage is observed as average (currently 57.8 per cent). Passengers should also be encouraged to be vigilant and assist in security upkeep. Security is one of the factors that would assist in making DART be seen as convenient, as Okagbue *et al.* (2015) observe in their study on motivation and challenges faced by commuters using BRT in Lagos, Nigeria.

4.3.9 ARRIVING WITHIN PLANNED TIME (Q3)

This attribute has a mean score of 2.72, indicating a positive score though not very high. This is likely to result from passengers spending an elongated period at the bus stations. Thus, passengers do not get on board in time because buses are overcrowded or there few buses on service, hence elongating the waiting time. As exhibited by Okgabue's *et al.* (2015) study on "*Motivation and challenges faced by commuters using BRT in Lagos, Nigeria,*" more buses should be availed to reduce total commuting time. Amiegbebhor and Dickson (2014), studied the impact of bus rapid transit on commuter satisfaction in Lagos state, Nigeria, and found out that inadequate number of buses constituted a major problem. In this regard, the introduction of bus scheduling and timing may assist in reducing the waiting time for commuters at DART stations and terminals (ibid.).

4.3.10 FAILURE TO GET ON BOARD (Q22)

This attribute has an inverted mean score of 2.88 (actual score is 3.12). This score is positive but very close to the median hence indicating that there is considerable failure of passengers to get on board the buses while at the stations. Failure to get onto the buses tends to affect the passengers' planned time of arrival. This failure could be a result of both overcrowding and disorder while entering the buses or passengers spending a long period at ticket booths.

4.3.11 COMFORT INSIDE THE BUSES (Q2)

This attribute has a mean score of 3.29. This is the highest figure above median, representing the lowest satisfaction of all attributes. Thus, comfort inside the buses is a major concern among passengers interviewed. Buses overcrowding is one of the reasons behind this situation. In a study done by Batarce *et al.* (2015) on public transport in Santiago, Chile, overcrowding was also cited as one of the sources of discomfort. Overcrowding may prevent non-captive passengers from shifting to BRT and, thus, frustrating the efforts aimed at reducing road congestion and motor gas emissions (Batarce *et al.*, 2015). This item needs attention if DART is to attract more choice passengers and assist in reducing road congestion with the city currently characterised with traffic jams.

4.4 Means of participants' perceptions on service quality attributes with respect to gender and age

Means of service quality attributes with respect to gender and age have been presented in Tables 8 and 9.

Table 8: Mean distribution of service quality attributes with respect to gender

Question	Attributes	Mean	
		Male	Female
1	Affordability of fares	2.85	2.77
2	Comfort inside the buses	3.35	3.23
3	Arriving within planned time	2.70	2.74
4	Frequency in relation to needs	2.64	2.46
5	Convenience of service hours	2.38	2.32
6	Time to reach destination	2.61	2.60
8	Buses breakdown	4.27	4.32
9	Adequacy of security at stations	2.80	2.71
10	Adequacy of shelter	2.64	2.61
17	Frequency of use of services	2.42	2.36
22	Failure to get on board as buses pass	3.16	3.08

Results summarized in Table 10 show that there is no difference in the perception of positive quality attributes between male and female respondents.

Table 9: Mean distribution of service quality attributes with respect to age groups.

Question	Attributes	Age 18-25	26-35	36-45	>45
1	Affordability of fares	3.02*	2.84	2.69	2.56*
2	Comfort inside the buses	3.36	3.31	3.30	2.95**
3	Arriving within planned time	2.87*	2.70	2.70	2.32**
4	Frequency in relation to needs	2.53	2.59	2.57	2.37*
5	Convenience of service hours	2.28	2.42	2.34	2.32
6	Time to reach destination	2.65	2.66	2.53	2.47*
8	Buses breakdown	4.41	4.14	4.33	4.42
9	Adequacy of security at stations	2.71	2.93*	2.70	2.37*
10	Adequacy of shelter	2.67	2.70	2.59	2.32*
17	Frequency of use of services	2.49	2.44	2.21	2.47
22	Failure to get on board as buses pass	3.05	3.09	3.19	3.26

* Items showing difference in levels of satisfaction. (>5% difference from the mean)

** Items showing significant difference in levels of satisfaction (>10% from the mean)

Results in Table 9 show that there is a different perception in terms of positive perception of service quality attributes among age groups. The differences were noted in the age group of >45 years. These involved responses in Q1, Q2, Q3, Q4, Q6 and Q9 with Q2 and Q3 showing significant differences in perception.

4.5. Assessment of other matters that provide useful information in the scaling up of DART

In addition to service quality attributes presented in Table 6, the study considered and analysed other aspects of DART services with the purpose of obtaining information that may assist in understanding further the passengers' perception of DART services and thereby creating the basis for making better decisions aimed at improving the current phase while implementing the enhanced future phases. Table 10 presents the results:

Table 10: Results of responses to other questions including those not having direct impact on service quality but are necessary inputs in improving services

Question number	Attribute	N	(%)
12	**Did you have information about DART services before they started**		
	Yes	135	(52.7)
	No	121	(47.3)
13	**Were you involved in any way in contributing your opinion and views about the services**		
	None reported to have been involved due to lack of opportunity to do so.		
14	**How do you get to the DART station?**		
	By Daladala	88	(34.5)
	By walking	90	(35.3)
	By Bodaboda (motor-cycle)	44	(17.3)
	By Bajaji (tri-motorcycle)	15	(5.9)
	By Own car, Taxi hire (e.g. Uber)	18	(7.0)
15	**How long does it take you to reach the DART station?**		
	Up to 10 minutes	78	(32.2)
	11-20 minutes	88	(36.4)
	21 – 30 minutes	40	(16.5)
	31 – 40 minutes	6	(2.5)
	Beyond 40 minutes	17	(7.0)
	Varies	13	(5.4)
16	**How long does it take to wait for the bus?**		
	Up to 10 minutes	105	(41.5)
	11 – 20 minutes	98	(38.7)
	21 – 30 minutes	23	(9.1)
	31 – 40 minutes	6	(2.4)
	Beyond 40 minutes	21	(8.3)
20	**Purpose of trip**		
	Commuting to the workplace	106	(41.6)
	To School/College	52	(20.4)
	Shopping/leisure	22	(8.6)
	Place of worship	4	(1.6)
	Other income earning activities	71	(27.8)
24	**Challenges faced while using DART**		
	Congestion on the bus	63	(36.6)
	Congestion at bus stops	1	(0.6)
	Lack of service when it rains	8	(4.7)

Few buses	17	(9.9)
Long queue at ticket booth	31	(18)
Late arrival	2	(1.2)
Long waiting time	20	(11.6)
High fare	9	(5.2)
Theft	5	(2.9)
Scrambling while boarding	9	(5.2)
Lack of seats	7	(4.1)

The results summarised in Table 10 are discussed as follows:

4.5.1 INFORMATION ABOUT DART BEFORE SERVICES COMMENCEMENT (Q12)

About 47.3 per cent of the passengers interviewed did not have any information about DART before it commenced its services whereas 52.7 per cent had received information. None of the passengers interviewed had been involved in contributing ideas on DART operations. This could explain why inadequacy in items such as shelter is observed. If public opinions and ideas were sought and incorporated, some shortcomings might have been avoided (e.g. the scrambling while entering and alighting from buses observed). Prospective passengers would also have recommended the use of mobile phone money services for the purchase of tickets as Tanzania is one of the leading countries in the world in mobile money transfers (Tanzania Invest. 2018) and since ownership of mobile phones has been increasing with penetration reaching 83 per cent by March 2017 (Lancaster, 2017). The use of the exact no-change denominations also appeared as a problem as personal experience reveals that Daladala operators strongly insist on giving them the exact amounts when boarding buses.

4.5.2 GETTING TO DART STATION (Q14)

About 35.3 per cent of the passengers interviewed would get to DART stations on foot and 34.5 per cent use Daladala. Since more than one-third of DART users' access stations through the Daladala (privately owned mini-buses), it is important and necessary to integrate DART with the para-transit services.

4.5.3 TIME TAKEN TO REACH DART STATION (Q15)

The majority (68.6%) of DART users, who took part in the study, take less than 20 minutes to reach DART stations. In other words, more than half of the passengers interviewed spent less than 20 minutes to reach the DART stations.

4.5.4 BUSES WAITING TIME (Q16)

In their responses, 41.5 per cent of the DART users reported to have been spending up to 10 minutes to wait for the bus whereas 38.7 per cent spent between 11 and 20 minutes. This means that over 80 per cent of DART users spent less than 20 minutes waiting for buses. The 38.7 per cent who reported spending between 11 and 20 minutes could be the ones who cannot get on board and see buses pass while marooned at the stations (Q22), (17.9% reported "often" and 44.9% reported "once in a while") (Table 6). Therefore, it can be argued that with an increased number of buses, the waiting time for buses could be lowered to less than 10 minutes for many more users.

4.5.5 TRIPS PURPOSES (Q20)

41.6 Per cent of the passengers' interviewed used DART services to get to their workplaces, whereas 27.8 per cent use the services to go to other income generating activities, and 20.4 per cent use DART to get to schools and colleges. This means 69.4 per cent of DART users depend on DART services for their livelihoods. Therefore, DART services have a significant impact on the income of Dar es Salaam residents.

4.5.6 CHALLENGES FACED WHILE USING DART SERVICES (Q24)

The most significant challenges recorded amongst the DART users are congestion inside the buses (36.6%), long queues for tickets (18.0%), and long waiting time for the buses (11.6%), few buses (9.9%), and high fare (5.2%). Other challenges include scrambling while boarding (5.2%), lack of service when it rains (4.7%), lack of seats (4.1%), theft (2.9%), and congestion at the bus stops (0.9%). Thus, increasing the number of buses would reduce congestion on the buses; and an increase of frequency would reduce the waiting time. Ticket purchasing has to be improved to reduce long queues. Although lack of buses during rainy seasons may appear insignificant (as indicated by only 4.7 per cent (Table 7), it needs to be taken seriously as a step towards improving DART's uninterrupted services. Personal experience has revealed frequent stoppages of DART services partially or wholly due to infrastructure challenges, especially at the Jangwani stretch where trunk roads become submerged in floodwaters whenever there are heavy rains.

4.5.7 DART INFRASTRUCTURE IN RELATION TO OTHER DEVELOPMENTS

The researcher has also established that there was no effective co-ordination between new developments in the city and DART infrastructure. This means DART does not integrate fully the potential passenger catchment areas currently being planned or constructed. Indeed, areas of large concentration of people and potential collection of BRT passengers are not well linked to BRT terminals/stations. For instance, the planned phase III of the BRT (along Nyerere Road) does not seem to have a seamless link with Mwalimu Julius Nyerere international airport terminals (as evidently seen on the almost completed terminal 3). These terminals are expected to handle about 6.5 million passengers annually when the newly constructed Terminal 3 becomes fully operational (Construction review, 2017). Moreover, the Morocco terminal does not easily connect to the Morocco square shopping mall and mixed use facility currently under construction and expected to have gross floor area of 110,000sqm (National Housing Corporation Tanzania [NHCTZ], 2018). To address these shortcomings, integrated land-use and transportation planning should be adopted while developing further phases of BRT as the system can potentially affect land development (Deng &Nelson, 2011). As the city of Dar es Salaam continues to grow, and with only selected roads scheduled to be finally covered by the BRT infrastructure (Figure 1), parallel efforts should be made to improve para-transit services so that feeder and distributor services are efficient. Furthermore, road infrastructure, as highlighted by Nhundu (2013), should be improved and appropriate links to BRT terminals and other areas of passengers' concentration should be planned.

CHAPTER FIVE

5.0 DISCUSSION

This chapter discusses the findings in accordance with the specific objectives of the study (as described in item 1.3.2). The findings discussed are in relation to the responses received from the respondents. The chapter also discusses the mean score of the attributes. Then it summarises the challenges that DART users' face and the measures they suggest as essential in bringing about the desired improvements.

5.1. Commuters' perception of service quality attributes (percentage of positive responses)

The study methodology developed questions with different response statements requiring respondents to select their perceptions on a 5-point Likert scale. The focus of each question is described in 4.2. Issues that merit discussion revolve around the interpretation of the perception of the respondents that selected the mid-point on the 5-point scale. In other words, is it neutral or does the wording of some questions allow the responses to represent a positive or a negative value. For all the questions except questions 8 (bus breakdowns) and 22 (not being able to get on board) a score of 1 is considered as the most positive. For questions 8 and 22, a score of 5 is considered as the most positive and the mean scores of 4.32 and 3.12 can be reversed to 1.68 and 2.88 respectively. Commuters' perception of the value of the existing services is determined by a combination of attributes as presented in Table 11 and further reflected in Figure 15.

Table 11 Percentage distribution of positive scores in relation to the questions

Question	Question statement	Positive perception		Score (%)
		From	To	
Q19	Continuing using DART services	*		96.9
Q9	Adequacy of security at stations	Average	Very adequate	91.4
Q6	Time to reach destination	Average	Very short	88.3
Q10	Adequacy of shelter	Average	Very adequate	87.6
Q7	Safe to enter/exit buses	*		81.7
Q13	Universal access	*		81.1
Q22	Failure to get on board buses	Once in a while	Very less often	78.1
Q5	Convenience of service hours	Convenient	Very convenient	77.4
Q17	Frequency of use of services	Often	Very often	73.4
Q1	Affordability of fares	Average	Very affordable	72.7
Q8	Buses breakdown	Less often	Very less often	68.8
Q11	Easiness of buying tickets	*		64.6
Q4	Frequency in relation to needs	Satisfactory	Highly satisfactory	63.5
Q2	Comfort inside buses	Average	Very comfortable	45.0
Q3	Arriving within scheduled time	Most of times	Always	43.2

- Positive perception refers to "Yes" response.

Determination of positive or negative response follows a discussion from 4.2.1 to 4.2.16.

The average percentage of scores of the recorded positive values is 74.2 per cent. This average does not differ much from the general perception of DART services as Table 5 illustrates with 77.0 per cent of the passengers' interviewed considering the services as satisfactory ("average" was considered as positive). These observations tally with the average mean scores for questions in Table 7 and Figure 15 (inverted scores for Q8 and Q22). The average of means stands at 2.58, which is on the positive side of the scale, was used.

The majority (96.9 per cent) of the DART users said they would continue using services (Q19). This observation is related to the fact that 80.6 per cent of the interviewed passengers are captive passengers (Q18) and that 88.3 per cent find time to reach destination as acceptable (Q6). Positive perception on travel time seems to override the lower perception on arriving within planned time (perceived positive by 43.2 per cent [Q3]) (Figure 13).

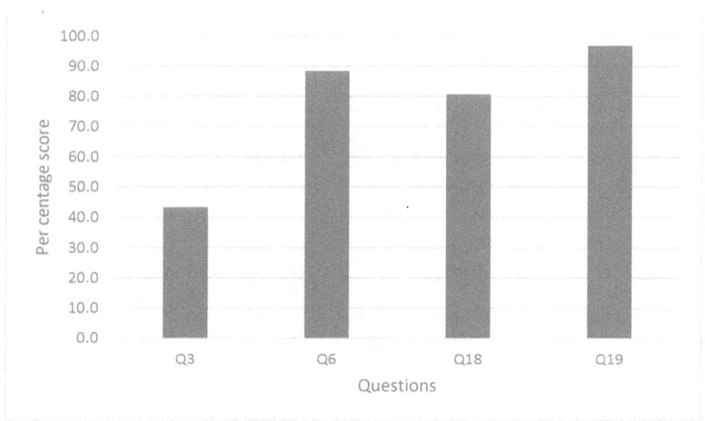

Figure 13: Positive response for questions 3, 6, 18 and 19 compared

These attributes have been compared, as they are interrelated. Captive passengers are more likely to continue using DART services because the time it takes to reach their destination is acceptable to them despite the discomfort they experience while riding on the buses and despite the failure of arriving within the scheduled time (Q3). The same negative perceptions are probably felt while using Daladalas. This continuance of using DART services is also related to fact that 77.4 per cent of the passengers find service hours as convenient (Q5). Moreover, 72.7 per cent of the users find the fares charged as affordable (Q1) and that 63.5 per cent of the users' travel needs are met by the frequency of services (Q4), which is related to the finding that 73.4 per cent of the passengers interviewed are frequent users of DART services (Q17) (Figure 14).

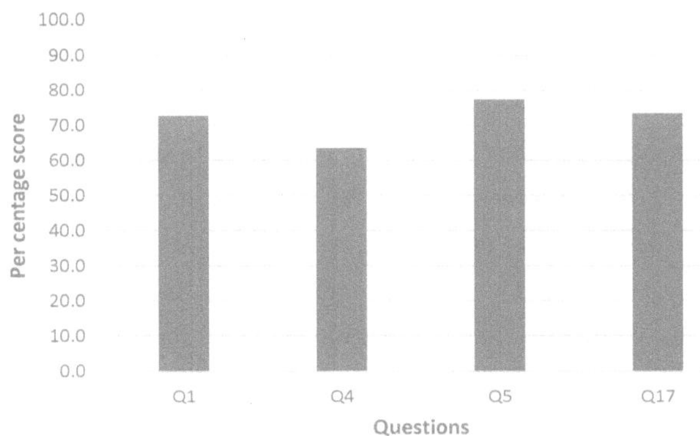

Figure 14: Positive responses for questions 1, 4, 5 and 17 compared

These attributes have been compared since the frequency of use is due to the convenience passengers experienced. Q1, Q4, and Q5 are the factors determining the public convenience of BRT.

5.2 Commuters' perception of service quality attributes (Means of positive responses)

Means of positive responses have been presented graphically for illustration of general perception of service quality attributes.

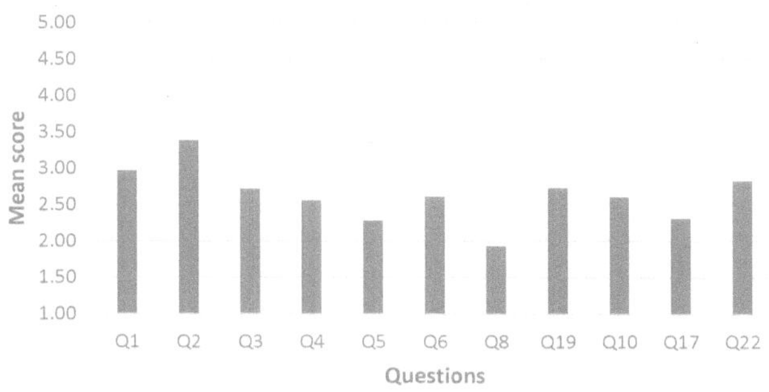

Figure 15: Mean distribution of DART passengers' responses to service quality questions

Scores for Q8 and Q22 have been inverted for easy comparison with the rest.

Figure 15 shows that Q2 has the highest value (3.29) indicating that this attribute (comfort inside the buses [Q2]) is the most negatively perceived attribute; meaning DART users' satisfaction with the comfort found inside the buses is minimal. The figure also demonstrates that affordability of fares (Q1) and failure to get on board the buses (Q22) remain a matter of concern to DART users, as these attributes have not been felt highly positive by DART users. On the other hand, Figure 15 indicates that breakdowns of DART buses were rare occurrences, hence not featuring as a serious matter of concern among the riders. The mean scores for Q8 and Q22 have been reversed to facilitate comparison (as shown in Table 7).

5.3 Passengers' perception on shortfalls and challenges encountered while using the DART system

Passengers' perception on the shortfalls and challenges encountered have been summarised and presented in Table 12. These are responses to questions 2, 3, 7, 10, and 11 of the questionnaire. The table also provides measures suggested by the passengers towards overcoming the DART challenges.

Table 12: Shortfalls, reasons, challenges and measures suggested by passengers

Question number	Attribute/shortfall	Number of reasons given	Reason for the shortfall (*)	Suggested measure
2	Comfort on the buses	3	Too many passengers compared to buses available (17) Congestion in a single trip (63) Others such as theft, structures issues for shorter people, etc.	To increase the number of buses To limit the number of passengers per trip To install seats in existing buses
3	Never arriving on time at a specified destination	1	Long waiting times at the bus station (20)	There should be a specific timetable at all times of the day
7	Unsafe to enter/exit the bus	3	Too many passengers at the entry and exit (9) Few doors in relation to the congestion (5) Associated thefts (5)	**NOT ASKED**
10	Inadequacy of shelter at the DART stations	2	Lack of sitting facilities (15) Few ticketing windows (30)	Sitting structures and more ticketing windows should be put in place.
11	Not easy to buy DART tickets	3	Very long queues due to too many passengers (31) Lack of change (balance in case you give a high valued money) (1)	**NOT ASKED**

*** ( ) Number of responses shown in brackets**

The information in Table 12 indicates that congestion on buses appears as a major source of discomfort on DART buses. The suggested measures were not asked for items (7) and (11) as they are found to require a more technical intervention which users may not have been able to suggest. For instance having too many passengers at exit may pose a danger for riders to be injured due to significant wide horizontal gap between the buses and the platform. Measures would include optical guidance systems, alignment markers, or boarding bridges (ITDP, 2016).

5.4 Incorporation of public views in designing future DART phases

The study has indicated that 52.7 per cent of the respondents had information about DART services prior to the introduction of the services, whereas, 47.3 per cent did not have such information (Table 10). All the respondents indicated not to have been involved in any way during the design and planning stages of the DART project. The main reason for this non-inclusion was cited as lack of opportunity for them to do so. This finding contrasts with experiences in other BRT projects. Kumar *et al.* (2012) in their study on international experience in Bus Rapid Transit (BRT) implementation in which they synthesised lessons learned from Lagos (and Johannesburg, Jakarta, Delhi and Ahmedabad) found a strong communication programme as one of the factors that contributed to the success of BRT Lite (Lagos). The programme made use of TV, radios, and websites and enhanced communication amongst many public groups and stakeholders. As a result, stakeholders were made aware of the project plans and benefits. This two-way approach brought about meaningful contribution to the development of BRT Lite (Lagos).

CHAPTER SIX

6.1 CONCLUSIONS

The majority of DART users are generally satisfied with the services provided. For DART services to attract wider usage efforts should be made to increase satisfaction level beyond average. The following are the main conclusions from the study:

a) As 55 per cent of the passengers interviewed found the buses as uncomfortable, this is an area requiring critical attention by DART when rolling out future phases and when embarking on services improvement in the current phase. In addition, one-third of the DART users found the frequency in relation to their travel needs as unsatisfactory. This aspect requires more work to ensure that the necessary improvements are made as frequency determines convenience.

b) The fares charged should be made more affordable. In this regard, 27.3 per cent of the users interviewed currently consider DART services as unaffordable; as such, there is need to make the services affordable to a broad-based clientele for them to keep on using DART services.

c) In the study, 91.4 per cent of the DART passengers found security at the stations as acceptable such that passengers feel secure while they are on these stations. This positive state should be maintained by ensuring that passengers continue feeling safe and that passengers should report any suspicious incidences that may hamper security.

d) Furthermore, 87.6 per cent of DART passengers found shelters at the stations as adequate. The implication is that passengers were comfortable before boarding the buses, the shortfalls observed such as lack of sitting facilities at the trunk stations notwithstanding.

e) In addition, 88.3 per cent of the DART passengers reported being satisfied with travel time but more than half fail to arrive at their respective destinations within the planned time (56.7%). The positive observation on travel time should be maintained. Passengers' journey planning should be assisted by increasing the number of buses to cope with possible increase of the number of commuters. Unplanned obstructions such as traffic police manually controlling the traffic flow during peak hours (observed by the researcher frequently) when the traffic lights were functioning should be largely avoided to support BRT unique character of right of way through most of the road intersections. Furthermore, timetables should be provided to assist passengers in planning their journeys.

f) Critical shortfalls perceived by the passengers included discomfort while on the buses and their inability to board buses as they pass through stations; because the buses are already overloaded.

g) There was insufficient communication made to the public about the DART project during the pre-planning and designing stage with the aim of incorporating public views in the eventual project.

6.2 LIMITATIONS OF THE STUDY

The results of this study were primarily based on DART as one system. The study did not distinguish between DART as an agency and UDART as operators currently offering the BRT services, hence operational constraints did not form part of the study. Moreover, the study did not make any comparison with other public transport alternatives. Furthermore, this study did not include issues related to urban dynamics and sprawling pattern of the city of Dar es Salaam.

The following difficulties were experienced during the study:

6.2.1   DIFFICULTIES IN DOING SURVEYS

a) Some passengers interviewed did not provide responses to all the questions in the questionnaire.

b) Taking notes during peak hours was difficult due to overcrowding in the buses.

6.2.2 WORDING DIFFICULTIES

Wording in some questions did not provide definite responses. For instance, Q13 (about universal access) the question should have been specific on whether access to the groups of physically challenged users was provided (sight, movement, and hearing). The purpose of the trip (Q20) should have been given as options for the respondents to choose rather than leaving them to provide answers based on their interpretation. In Q7 (Safe enter/exit) safety should have elaborated whether it was in relation to theft or being hurt.

6.2.3 DIFFICULTY IN SCORES INTERPRETATION

Some difficulties were experienced while interpreting scores. This arose from the way questions were asked. For two questions, a higher score meant a positive perception whereas for others it meant a negative perception. This necessitated inverting scores (see, for example, Questions 8 and 22) to facilitate the comparison of scores.

6.2.4   MEASURING THE IMPORTANCE OF SERVICE ATTRIBUTES

While the study was only intended to determine the levels of satisfaction among DART passengers, providing the respondents with the opportunity to rank the different attributes of DART would have added another dimension when prioritizing interventions of improve the satisfaction of DART passengers in future.

6.3 RECOMMENDATIONS

The following recommendations are put forward so that DART services quality as perceived by passengers may improve.

6.3.1 RECOMMENDATIONS TO DART AUTHORITY

a) The number of buses should be increased. The number of passengers transported by DART system per day should be used to determine the appropriate number of buses needed by taking into account the limit of congestion that would guarantee passengers' comfort. As recommended by Amiegbebhor and Dickson (2014) in the case of Lagos, displaying the seating and standing capacity on the buses for passengers' information and enforcing compliance may assist in reducing overcrowding.

b) DART should consider providing timetables for services to enable users to plan for their trips accordingly.

c) Through public education on the use of DART facilities, an orderly entry and exit of the buses should be promoted. Users should be discouraged from scrambling onto the buses and should be aided to develop a culture of orderly entry while providing more buses to ease the problem of scrambling.

d) Travel cards should be made available while discouraging the current paper ticket system. Topping up of cards by using mobile services such as Mpesa, Tigopesa, and Airtel Money should be enhanced.

e) As ownership of mobile phones in the country increases, development of a mobile phone based system that can assist in paying and accessing DART services should be considered.

f) There is a need of increasing the number of ticket vending stations and consider having them in other places outside the DART stations.

g) Public views should be incorporated in the design of future phases as the appropriateness of the resultant system impacts on the quality of transport service (Deng & Nelson, 2011). Public views should be incorporated beyond the construction stage from time to time so that further improvement of DART services would take into account these views and, thus, let the public own the system, or feel to be an integral part of it.

h)   DART should encourage operators to use appropriate bus maintenance plans so that the low level of buses breakdowns observed can be sustained.

i)   Integrated land-use and transportation planning should be adopted while developing further phases of BRT as the system can potentially affect land development. Seamless connection between DART stations and areas of potential passenger concentration (such as Terminal 3 of Mwalimu Julius Nyerere International Airport and Morocco square mixed-use facility) should be considered.

6.3.2   RECOMMENDATIONS TO DART SERVICES USERS

These recommendations should be incorporated into any existing educational or awareness programme so that they reach DART users regularly:

a)   Passengers opting to pay by cash at the ticket vending stations should be encouraged to come with exact amounts and abstain from use high denominations currency.

b)   Passengers should refrain from scrambling to get onto the buses and to alight from them, as this will guarantee safety for all the users.

c)   Passengers should be encouraged to be vigilant and report any suspicious bad behaviour at the stations or on the buses.

6.4 SUGGESTIONS FOR FURTHER STUDY

a) A study on importance of service quality attributes

This study focused on the level of satisfaction of service quality attributes and did not include ranking of the importance of attributes. It is therefore suggested that any future study should include ranking of the importance of attributes. The result of such ranking would assist when prioritizing interventions aiming at improving DART services.

b) DART service quality in relation to the urban sprawl pattern of the city of Dar es Salaam

The impact of uncontrolled urban sprawl on DART services should be studied. Since the city continue to expand the needs for transport services beyond the current end stations would increases; thus, it may be necessary to establish whether to limit city expansion or to continue expanding BRT services.

c) Impact of service quality and reliability of para-transit operators in relation to DART services

BRT alone is not the panacea to all the urban transport problems (Arrive alive, 2017); thus, the impact of other service providers that feed the DART system should be studied. This study has established that nearly a third of passengers using DART system use Daladala services to reach the stations. Thus, poor services provided have a negative impact on the overall journeys made by passengers.

d) Security matters

The security attributes measured in this study considered security at the stations only. It is suggested that security within the last mile be included in the future studies to give a clearer picture of security of DART users. Security within the last miles will include passages from feeder stations to trunk stations which affect passenger security.

e) Appropriateness of the buses and infrastructure in DART phase I

The buses in use are part of the pilot phase. A study should be made to establish how appropriate they are and to what extent they have contributed to the passengers' discomfort. Critical BRT infrastructure design matters and challenges the staff face in delivering their services should also be studied in future.

REFERENCES

Abdullah, K., Jan, M. T., & Manaf, N. H. A. (2012). A structural equation modelling approach to validate the dimensions of SERVPERF in airline industry of Malaysia. *International Journal of Engineering and Management Sciences, 3*(2), 134 141.

Adebambo,S. & Adebayo, I., (2009). 'Impact of bus rapid transit system (BRT) on passengers' satisfaction in Lagos Metropolis, Nigeria', *International Journal of Creativity and Technical Development,* 1(3): 106-119.

Afolabi, O. (2016), Commuter perception and preferences on the bus rapid transit in Lagos state, JORIND 14(2) available at www.transcampus.org/journal, 34-40.

Allen, H., (2013), Africa's First Full Rapid Bus System: the ReyaVaya bus system in Johannesburg, Republic of South Africa: A case study prepared for global report on human settlements 2013. Available from http://www.unhabitat.org/grhs/2013

American Public Transport Association (APTA) (2010), Bus rapid transit service design, APTA BTS-BRT-RP-004-10.

Amiegbebhor, D., &. Dickson, O.F. (2014), 'Impact assessment of bus rapid transit on commuter satisfaction in Lagos state, Nigeria', *JORIND* 12 (2):198 – 208.

Arrive alive, (2017), www.arrivealive.co.za, accessed on 17th October 2017.

Batarce, M. Munoz, J. Ortuzar J. Raveau, S. Mojica, C. & Rios, R. (2015), Evaluation of passenger comfort in bus rapid transit systems, *Technical note No. IDB-TN-770, Inter-American Development Bank, Infrastructure and Environment Sector, Transport Division,* available at https://publications.iadb.org

Bertam, D., (2017), Topic report available at
http://scholar.google.com/scholar_url?url=https%3A%2F%2Fwww.researchgate.net
%2Fprofile%2FFarhad_Waseel%2Fpost%2FHow_can_one_help_people_create_sim
ple_scales_from_Likert-
scored_items%2Fattachment%2F5c9e2e9acfe4a7299499df34%2FAS%253A741800

755068929%25401553870490367%2Fdownload%2Ftopic-dane-likert.pdf&hl=en&sa=T&oi=ggp&ct=res&cd=0&d=62654880095531585&ei=8gCs XcO_N8qpmQHapr_gCg&scisig=AAGBfm0fHuyWkpo3QM0wxpsxMdpX9x8NRg &nossl=1&ws=1216x567&at=Likert%20scales&bn=1. Accessed on 20th October 2019.

Bruun, E., Del Mistro, R., Venter, Y. & Mfinanga, D., (2015), The state of public transport in three Sub-Saharan African cities. In: Behrens, R. McCormick, D. & Mfinanga, D. ed. *Paratransit in African cities*, Abingdon: Routledge, 26 – 58.

Cain, A., & Flynn, J., (2013), Examining the ridership attraction potential of bus rapid transit: A quantitative analysis of image and perception, *Journal of public transport*, 16 (4): 63 -82.

Chou, P.-F., Lu, C.-S., & Chang, Y.-H. (2014). Effects of service quality and ustomer satisfaction on customer loyalty in high-speed rail services in Taiwan. *Transportmetrica A: Transport Science*, *10*(10), 917 945. http://doi.org/10.1080/23249935.2014.915247

CITISCOPE (2017), http://citiscope.org/story/2017/new-bus-rapid-transit-system-earns-dar-es-salaam-2018-sustainable-transit-award. Accessed on 2nd November 2017

Construction Review (2017), https://constructionreviewonline.com/2017/12/julius-nyerere-international-airport-terminal-three-67-complete/. Accessed on 3rd August 2018

Dar es Salaam Rapid Transit (DART), (2014), Project information memorandum.

Del Mistro, R. & Behrens, R. (2015), Integrating the informal with the formal: An estimation of the impacts of a shift from paratransit line-haul to feeder service provision in Cape Town, Case Studies on Transport Policy 3, 271-277

Deng T., & Nelson, J., (2011), Recent Development in Bus Rapid Transit: A Review of the Literature, *Transport Reviews*, 31 (1) 69-96, DOI 10.1080/01441647.2010.492455

Deng, T., & Nelson, J., (2012), The perception of bus rapid transit: A passenger survey from Beijing southern axis BRT line 1, *Journal of transportation planning and technology*, 35 (2): 201-219.

Dhanavandan, S., (2016), Application of Garrett ranking technique: Practical approach, International journal of library and information studies, Vol. 6(3), pages 135-140. Available at http://ijlis.org/img/2016_Vol_6_Issue_3/135-140.pdf

Eboli. L., & Mazulla. G., (2007), Service quality attributes affecting customer satisfaction for bus transit. *Journal of Public Transportation*, 10 (3): 21 – 34.

Entrepreneur (2017), https://www.entrepreneur.com/encyclopedia/branding. Accessed on 7th November 2017.

Etikan, I., Musa, S., & Alkassim, R., (2016), Comparison of Convenience Sampling and Purposive sampling. *American Journal of Theoretical and Applied Statistics.* 5 (1): 1-4. Doi: 10.11648/j.ajtas.20160501.11

Fick, G. R., & Ritchie, J. R. B. 1991. Measuring Service Quality in the Travel and Tourism Industry. *Journal of Travel Research*, *30*(2), 2 9. http://doi.org/10.1177/004728759103000201

Government of United Republic of Tanzania (1997), Executive Agency Act No. 30 of 1977

Government of United Republic of Tanzania (1982), Local Government (Urban Authorities) Act, No. 8

Government of United Republic of Tanzania (2007), Road Act, No. 13 of 2007.

Government of United Republic of Tanzania (1973), Road Traffic Act, No. 13 of 1973.

Government of United Republic of Tanzania (2008), Standards Act of 2008

Government of United Republic of Tanzania (2001), Surface and Marine Transport Regulatory Authority Act, 2001.

Government of United Republic of Tanzania (1973), Transport Licensing Act No. 1 (CAP 317) of 1973.

Hussein, A. & Hapsari, R., (2014), How quality, value and satisfaction create passenger loyalty: An empirical study on Indonesia bus rapid transit passenger, *The International Journal of Accounting and Business Society*, 22 (2): 95-115.

International Association of Public Transport (UITP), (2010), *Symposium on Public Transportation in Indian Cities with special focus on bus rapid transit (BRT) system*, UITP, Delhi.

ITDP (2012), The BRT Standard

ITDP (2016), The BRT Standard

Japan International Cooperation Agency, (2008), Dar es Salaam transport policy & system development master plan-Technical Report.

Klein, M., (2014), Customer satisfaction is not customer loyalty, Accessed 1st February 2017, from http://www.dmnews.com/customers-satisfaction-is-not-customer-loyalty-/article/313592/

Kothari, C., (2009) *Research Methodology: Methods and Techniques*, Delhi: New Age International (P) Limited

Kumar, A. Zimmerman, S. & Agarwal, O. (2012), International experience in Bus Rapid Transit (BRT) implementation: Synthesis of lessons learned from Lagos, Johannesburg, Jakarta, Delhi and Ahmedabad. Available at https://openknowledge.worldbank.org/handle/10986/13049

Lai, W. T., & Chen, C. F., (2011). Behavioral intentions of public transit passengers The roles of service quality, perceived value, satisfaction and involvement. *Transport Policy, 18*(2), 318 325. http://doi.org/10.1016/j.tranpol.2010.09.003

Lancaster. H., (2017): Tanzania Telecoms, mobile, broadband and digital media – Statistics and Analyses, Buddecomm report available through https://www.tanzaniainvest.com Accessed on 3rd August 2018

Mahmoud, M., Verdinejad, F., Jandaghi, G. & Mughari, A., (2010), Analysis and establishment of bus rapid transit (BRT) on customer satisfaction in Teheran, *African Journal of Business Management,* 4(12): 2514-2518, available at http://www.academicjournals.org/AJBM

Mehta, P. (2011), Profile and Perception of Investors towards mutual funds-A study of selected cities of Gujarat state, Doctor of philosophy in commerce degree book, Veer Namad South Gujarat University, Surat

Moylan, E., Clonts, K., Wang, D., O'Connor, S., & Lounsbury, D. (2013). 'Public perception of bus rapid transit in the San Francisco Bay area' 13th World Conference on Transport Research, Rio de Janeiro, Brazil

Morton, C, Caulfield, B and Anable, (2016) Customer perceptions of quality of service in public transport: Evidence for bus transit in Scotland. Case Studies on Transport Policy, 4 (3). pp. 199-207. ISSN 2213-624X

National Housing Corporation Tanzania (2018),http://www.nhctz.com/moroccosquare/ Accessed on 3rd August 2018.

Nhundu, E., (2013), Challenges facing transport sector in providing quality service to the society: Case of public transport sector in Dar es Salaam (book), Open University of Tanzania.

Nkurunziza, A., Zuidgeest, M., Brussel, M., & Bosch, F., (2012a), Spatial variation of Transit Service Quality Preference in Dar es Salaam, *Journal of Transport Geography*, 24: 12-21, available at www.sciencedirect.com

Nkurunziza, A., Zuidgesst, M., Brussel, M., & Maarseveen, M., (2012b), Modelling Commuter Preferences for the Proposed Bus Rapid Transit in Dar es Salaam, *Journal of Public Transportation*, 15(2): 95-116. DOI: http://dx.doi.org/10.5038/2375-0901.15.2.5, available at http://scholarcommons.usf.edu/jpt/vol15/iss2/5

Okagbue, H., Adamu, M., Iyase, S., & Owoloko, E., (2015), On The Motivation and Challenges Faced by Commuters Using Bus Rapid Transit in Lagos, Nigeria, *The Social Sciences*, 10 (6): 696-701.

Olikova, I., (2016), Evaluation of quality public transport criteria in terms of passenger satisfaction, Transport and telecommunication, Vol 17, no 1, 18-27

Pakdil,F., & Aydin, Ö. (2007) Expectations and perceptions in airline services: An analysis using weighted SERVQUAL scores. *Journal of Air Transport Management*, *13*(4), 229 237. http://doi.org/10.1016/j.jairtraman.2007.04.001

Patel, B., & Makwana, T., (2015), A Study on Future Preferences & Satisfaction of BRTS Customers with Special Reference to Ahmedabad City, *International Journal of Scientific Research*, 4 (6): 527 - 529.

Potogvkina, N.,& Ugay, S., (2013), Quality assessment of transport service of passengers in Vladivostok, Russia, World applied sciences journal 24 (6): 809-813, 2013

Purcher. J., Park. H. & Kim. H.M., (2006). Public Transport Reforms in Seoul: Innovations Motivated by Funding Crisis. *Journal of Public Transportation*, 8 (5): 41 – 46

Redman, L., Friman, M., Garling, T& Hartig, T., (2013), Quality attributes of public transport that attract car users: A research review, Transport policy, vol. 25, January 2013, Pages 119-127

Steinem, T. Fukuda, A. Oshima, R. (2006), A study on the introduction of bus rapid transit system in Asian developing cities – A case study on Bangkok metropolitan administration project, *IATSS Research*, 30 (2): 59 – 69.

Sivakumar, T., Yabe, T., Okamura, T. & Nakamura, F., (2006), Survey design to grasp and compare user's attitudes on bus rapid transit (BRT) in developing countries, *IATSS Research*, 30 (2): 51 - 58.

Surface and Marine Transport Regulatory Authority (SUMATRA), (2011), Study on the user needs and management of public transport services in Dar es Salaam. Accessed through www.sumatra.go.tz/index on 15th January 2017.

Tanzania Invest (2018) https://www.tanzaniainvest.com/telecoms/mobile-money-transactions-2017-2018. Accessed on 27th November 2018

Ugo, P.D., (2014), 'The bus rapid system: A service quality dimension of commuter uptake in Cape Town, South Africa', *Journal of Transport and Supply Chain Management* 8(1), Art. #145, 10 pages. http://dx.doi.org/10.4102/jtscm. V8i1.145

World Bank (2013), http://www.worldbank.org/en/news/press-release/2013/01/15/additional-financing-tanzania-bus-rapid-transit-system-benefit-300000-commuters-create-80000-jobs. Accessed on 2nd November 2017.

Xiyuan, L., (2014), Bus Rapid Transit quality Analysis: A Case study of Hefei. (book). University of Hong Kong, Pokfulam, Hong Kong SAR. Retrieved from http://dx.doi.org/10.5353/th_b5479495.

Zeithaml, V.A., Bitner, M.J., & Gremler, D.D., (2006), Services Marketing: Integrating customers focus across firm (Fourth Edition), New York, United States of America, The McGraw-Hill Companies, Inc.

Ziyari K., & Hajisharifi, A., (2013), Investigate the level of Satisfaction of BRT system (Case Study: line (3) Terminal of ELM-O-SANAT-KHAVARAN), *Journal of Spatial Planning*, 3 (1; 8): 67-74.

APPENDIX I: Questionnaire (English version)

Questionnaire No.	
Date	
Name of Interviewer	

Good morning/afternoon,

My name is Shabani Mwatawala. We are doing an assessment of satisfaction of passengers using DART services. The information obtained will help to inform policy makers and implementers about quality of service as perceived by customers and will help them make appropriate changes to improve quality of services. It will only take few minutes. We will appreciate it if you could assist us.

1. How affordable do you find the fares charged

Very affordable	Affordable	Average	Not affordable	Highly unaffordable

2. How comfortable are the buses used

Very comfortable	Comfortable	Not sure	Uncomfortable	Very uncomfortable

If VERY UNCOMFORTABLE what is the main reason?

………………………..………………………………

…………………………………………………………………………………………………

…………………………………………………………………………………………………

What should be done to improve

………………………………………………………………………....................…..............…

…………………………………………………………………………………………………

…………………………………………………………………………………………………

3. How often do you arrive at your destination within planned time?

Always	Most of times	Few times	Very few times	Never

If NEVER what do you think is the reason?

……………………………………………………………

…………………………………………………………………………………

…………………………………………………………………………………

What do you think should be done to improve?

……………………..………………………………………………..…………………………

…………………………………………………………………………………

…………………………………………………………………………………

4. How would you describe the services frequency considering your travel needs?

Highly satisfactory	Satisfactory	Not sure	Unsatisfactory	High unsatisfactory

5. How convenient do you find hours of services?

Very convenient	Convenient	Not sure	Inconvenient	Highly inconvenient

6. How do you rate the time you take to reach your destination

Very short	Short	Average	Long	Very Long

7. If it safe for you to enter/exit the buses

[ ] Yes

[ ] No

If NO what is the main problem you see?

....

....

....

8. How often you see buses break down?

Very often	Often	Not sure	Less often	Never seen

9. How adequate is security at stations?

Very adequate	Adequate	Average	Inadequate	Very inadequate

10. How adequate is the shelter?

Very adequate	Adequate	Average	Inadequate	Very inadequate

If INADEQUATE/VERY INADEQUATE what do you think is missing and should have been

provided..

....

....

....

11. Is it easy for you to buy DART ticket?

☐   Yes

☐   No

If NO what do you think is the reason?

....

....

....

12. Did you have information about DART services before they started?

☐   Yes

☐   No

If YES where you involved in any way in contributing your opinion and views about the services?

☐   Yes

☐   No

13. Does BRT infrastructure provide universal access?

☐   Yes

☐   No

If NO what is your major aspect of concern?

....

....

....

What should be done to improve?

....

....

....

14. From your home or workplace how do you get to DART

station..

....

....

....

15. How long does it take you to reach DART station?

....

....

....

How long does it take you to wait for the bus?

....

....

....

....

16. How often do you use DART services

Very often	Often	Once in a while	Less often	Very less often

If VERY LESS OFTEN what is the main reason?

....

....

....

....

17. Do you own a private car?

[ ] Yes

[ ] No

If YES what is the main reason for you to use DART services instead of using your private
car..
....
....

18. Do you see yourself continuing to use DART services?

[ ] Yes

[ ] No

IF NO what is the main reason
....
....
....
....

19. What is the purpose of your trip
....
....
....

20. How much is your average monthly expenditure in a month?
....
....
....

21. While at bus stops how often buses pass by and you cannot get on board?

Very often	Often	Once in a while	Less often	Very less often

22. Generally how would you rate the overall experience of using DART services

Very satisfactory	Satisfactory	Neutral	Unsatisfactory	Very unsatisfactory

23. What challenges do you face while using DART services

………………….....……………………………………………………..……………

…………………………………………………………………………………………

…………………………………………………………………………………………

…………………………….

24. Respondent demographic data

    i)   Age profile

- [ ] 18 - 25
- [ ] 26 - 35
- [ ] 36 - 45
- [ ] 46 and above

    ii)   Gender

- [ ] Male
- [ ] Female

25. DART Section

Kimara - Gerezani

Kimara - Kivukoni

Kimara - Morocco

Kivukoni - Gerezani

Kivukoni - Morocco

Morocco - Gerezani

Ubungo - Gerezani

Ubungo - Morocco

Ubungo - Kimara

Thank you for your participation.

APPENDIX II: Questionnaire (Kiswahili version)

Namba ya Dodoso	
Tarehe	
Jina la Mhoji	

Habari ya asubuhi/Mchana/Jioni,

Jina langu ni Shabani Mwatawala. Tunafanya tathmini ya kuridhika na huduma za MWENDOKASI miongoni mwa watumiaji. Taarifa tutakazopata zitasaidia watungasera na waendesha mradi kufanya maboresho na mabadiliko katika huduma, Mahojiano ni ya muda mfupi tu. Tutashukuru ikiwa utatoa ushirikiano.

1. Unaionaje nauli inayotozwa?

Nafuu sana	Nafuu	Wastani	Ghali	Ghali sana

2. Unaelezeaje starehe kwenye mabasi?

Starehe sana	Starehe	Sina Hakika	Hakuna starehe	Hakuna starehe hata kidogo

Kama HAKUNA STAREHE HATA KIDOGO unadhani sababu ni nini?

....

....

....

Nini Kifanyike ili kuboresha

....

....

....

....

3. Mara ngapi unawasili unakokwenda katika muda uliopanga?

Mara zote	Mara nyingi	Mara chache	Mara chache sana	Huwa hutokei

Kama huwa haitokei unadhani sababu ni nini?

....

....

....

Unadhani nini kifanyike ili kuboresha?

....

....

....

4. Unaelezeaje wingi wa safari kulingana na mahitaji yako?

Za kuridhisha sana	Za kuridhisha	Sina hakika	Siyo za kuridhisha	Haziridhish hata kidogo

5. Unoaonaje masaa ya huduma?

Yanafaa sana	Yanafaa	Sina hakika	Hayafai	Hayafai Kabisa

6. Unauelezeaje muda unaotumia kufika unapokwenda

Mfupi sana	Mfupi	Wastani	Mrefu	Mrefu sana

7. Je ni salama kuingia/Kutoka ndani ya basi ?

☐ Ndiyo

☐ Hapana

Kama HAPANA unaona tatizo ni nini?

....

....

....

Ni Mara ngapi unaona mabasi yameharibika ukiwa safarini?

Mara nyingi sana	Mara nyingi	Sina hakika	Mara chache	Sijawahi kuona

8. Unaonaje hali ya usalama kwenye vituo

Ni wakutosha sana	Wakutosha	Wastani	Hautoshi	Hakuna usalama

9.   Unaonaje majengo na mabanda ya vituo?

Yanajitosheleza sana	Yanajitosheleza	Wastani	Hayajitoshelezi	Hayajitoshelezi kabisa

Kama HAYAJITOSHELEZI/HAYAJITOSHELEZI KABISA nini unadhani kimekosa na kingepaswa kuwepo.

10.   Je ni rahisi kwako kupata tiketi ya MWENDOKASI?

☐   Ndiyo

☐   Hapana

Kama HAPANA unadhani sababu ni nini?

....

....

....

....

11.   Ulipata kuwa na taarifa za MWENDOKASI kabla ya huduma kuanza rasmi?

☐   Ndiyo

☐   Hapana

Kama NDIYO ulishiriki kwa njia yoyote kutoa maoni yako juu yake ?

12.   Miundombinu ya MWENDOKASI inatumika kirahisi na watu wote ?
(walemavu au wenye matatizo mengine)

☐   Ndiyo

☐   Hapana

Kama HAPANA nini tatizo kubwa ulionalo?

....
....
....

Nini kifanyike ili kuboresha?

....
....
....
....

13. Unafikaje kwenye kituo cha MWENDOKASI unapotoka nyumbani au
ofisini?..
....
....

14. Inakuchukua muda gani kufika? ....
....
....
....

15. Inachukua muda gani kusubiri basi? ....

....

....

....

16. Mara ngapi unatumia huduma za MWENDOKASI

Mara nyingi sana	Mara nyingi	Mara moja moja	Mara chache	Mara chache sana

Kama MARA CHACHE sana sababu ni

nini..

....

....

....

17. Je uamiliki chombo cha usafiri

☐ Ndiyo

☐ Hapana

Kama NDIYO nini sababu ya wewe kutumia huduma ya MWENDOKASI

....

....

....

18. Je unategemea kuendelea kutumia MWENDOKASI?

☐ Ndiyo

☐ Hapana

Kama hapana nini sababu ya msingi?

....

....

....

....

19. Nini madhumuni ya safari yako? (sababu za safari)

....

....

....

20. Je matumizi yako ya maisha kwa mwezi wastani ni kiasi gani?

....

....

....

....

21. Unapokua kituoni ni mara ngapi basi hupita bila wewe kupanda?

....

....

....

Mara nyingi sana	Mara nyingi	Mara moja moja	Mara chache	Mara chache sana

22. Kwa ujumla unazionaje huduma za MWENDOKASI

Zinaridhisha sana	zinaridhisha	Sina maoni	Haziridhishi	Haziridhishi kabisa

23. Changamoto gani unakutana nazo wakati wa kutumia huduma ya
MWENDOKASI?...
....
....
....

24. Taarifa za aliyehojiwa

i)   Umri

☐   18 - 25

☐   26 - 35

☐   36 - 45

☐   Miaka 46 na kuendelea

ii)   Jinsia

☐   Mume

☐   Mke

25. Safari

☐   Kimara - Gerezani

☐   Kimara - Kivukoni

☐   Kimara - Morocco

☐   Kivukoni - Gerezani

☐   Kivukoni - Morocco

☐   Morocco - Gerezani

☐   Ubungo - Gerezani

☐   Ubungo - Morocco

☐   Ubungo - Kimara

Nashukuru kwa ushirikiano.

APPENDIX III: EBE Ethical clearance

APPLICATION FORM

Application for Approval of Ethics in Research (EiR) Projects
Faculty of Engineering and the Built Environment, University of Cape Town

Please Note:

Any person planning to undertake research in the Faculty of Engineering and the Built Environment (EBE) at the University of Cape Town is required to complete this form before collecting or analysing data. The objective of submitting this application prior to embarking on research is to ensure that the highest ethical standards in research conducted under the auspices of the EBE Faculty are met. Please ensure that you have read, and understood the EBE Ethics in Research Handbook (available from the UCT EBE Research Ethics website) prior to completing this application form.

APPLICANT'S DETAILS	
Name of principal researcher, student or external applicant	SHABANI WALAD MWATAWALA
Department	CIVIL ENGINEERING
Preferred email address of applicant:	mwatawala@gmail.co.tz
If Student	Your Degree: e.g. MSc, PhD etc.	M Eng
	Credit Value of Research e.g. 60/120/180/360 etc.	60
	Name of Supervisor (if supervised)	ROMANO DEL MISTRO
If this is a research contract, indicate the source of funding/sponsorship	N/A
Project Title	ASSESSMENT OF PASSENGERS SATISFACTION IN BUS RAPID TRANSIT (BRT): THE CASE OF DAR ES SALAAM RAPID TRANSIT (DART)

I hereby undertake to carry out my research in such a way that:

- there is no apparent legal objection to the nature or the method of research; and
- the research will not compromise staff or students of the other respondents or the University;
- the stated objective will be achieved, and the findings will have a high degree of validity;
- limitations and alternative interpretations will be considered;
- the findings could be subject to peer review and publicly available; and
- I will comply with the conventions of copyright and avoid any practice that would constitute plagiarism.

SIGNED BY	Full name	Signature	Date
Principal Researcher/Student/External applicant	SHABANI W MWATAWALA	Signature Removed	19/12/2017

APPLICATION APPROVED BY	Full name	Signature	Date
Supervisor (where applicable)		Signature Removed	
HOD (or delegated nominee) Final authority for all applicants who have answered NO in all questions in Section 1, and for all undergraduate research (including Honours)	ALPHOSE ZINGONI	Signature Removed	22/03/2018
Chair: Faculty EiR Committee For applicants other than undergraduate students who have answered YES to any of the above questions.			

APPENDIX IV: DART letter of approval

UNITED REPUBLIC OF TANZANIA
PRESIDENT'S OFFICE
REGIONAL ADMINISTRATION AND LOCAL GOVERNMENT
DAR RAPID TRANSIT AGENCY
2nd Floor, LAPF Tower, Bagamoyo Road, P. O. Box 724, Dar es Salaam, Tanzania
Tel: +255 22 2700486/280 Fax: +255 22 2700603 E-mail: info@dart.go.tz Website: www.dart.go.tz

Ref. No. BA.32/194/01/30 ·   Tuesday, May 29, 2018

Shabani Mwatawala,
C/O PSM Architects Co. Ltd
P.O.Box 76862,
DAR ES SALAAM.

RE: REQUEST FOR PERMISSION TO CONDUCT A SURVEY ON PASSENGERS SATISFACTION

This is in reference to your letter referenced SM/UCT/DART/03/2018 of May 11, 2018 about the subject matter above.

We would like to inform you that the Agency has granted permission for you to proceed with the study of passenger satisfaction in the DART system effective from May 29, 2018 to June 30, 2018.

The Agency will appreciate to get the outcome of the research for improving the BRT services.

Thanks for your cooperation

Signature Removed

Eng. Ronald M. Lwakatare
CHIEF EXECUTIVE

Copy to: Regional Police Commander,
Dar es Salaam Special Zone,
S. L. P 9140,
DAR ES SALAAM.

Managing Director,
UDA Rapid Transit Agency,
P.O.Box 872,
DAR ES SALAAM.

Managing Director,
China Tanzania Security Co. Ltd,
Plot 47, Block 6 Kijitonyama,
P. O. Box 19944,
DAR ES SALAAM.

All correspondence should address to: Chief Executive   Page 1 of 1

ASSESSMENT OF PASSENGER'S SATISFACTION OF BUS RAPID TRANSIT (BRT):
THE CASE OF DAR ES SALAAM RAPID TRANSIT (DART)

Informed Consent Form

My name is **Shabani Mwatawala**

We are doing an assessment of satisfaction of passengers using DART services as part of requirement of fulfillment of Master of Philosophy Degree at **University of Cape Town, South Africa.**

In addition to that requirement, the assessment aims to assisting the government of United Republic of Tanzania to gather information that may help to make necessary changes in next phases so that they operate efficiently and serve a passenger like you better.

This assessment will involve you by responding to questions included in the questionnaire, and it will take few minutes.

You are being chosen as we choose randomly among DART services users. Your participation in this assessment is entirely voluntary and all information you give will be treated with anonymity.

We will appreciate if you could help us and in turn help the government improve DART services.